AF598897

CLASSIC GUNS OF THE WORLD SERIES

THE MAT-49 SUBMACHINE GUN

AND PRECEDING FRENCH SUBMACHINE GUN DESIGNS, INCLUDING THE MAS-35

LUC GUILLOU

THE MAT-49 SUBMACHINE GUN

AND PRECEDING FRENCH SUBMACHINE GUN DESIGNS, INCLUDING THE MAS-35

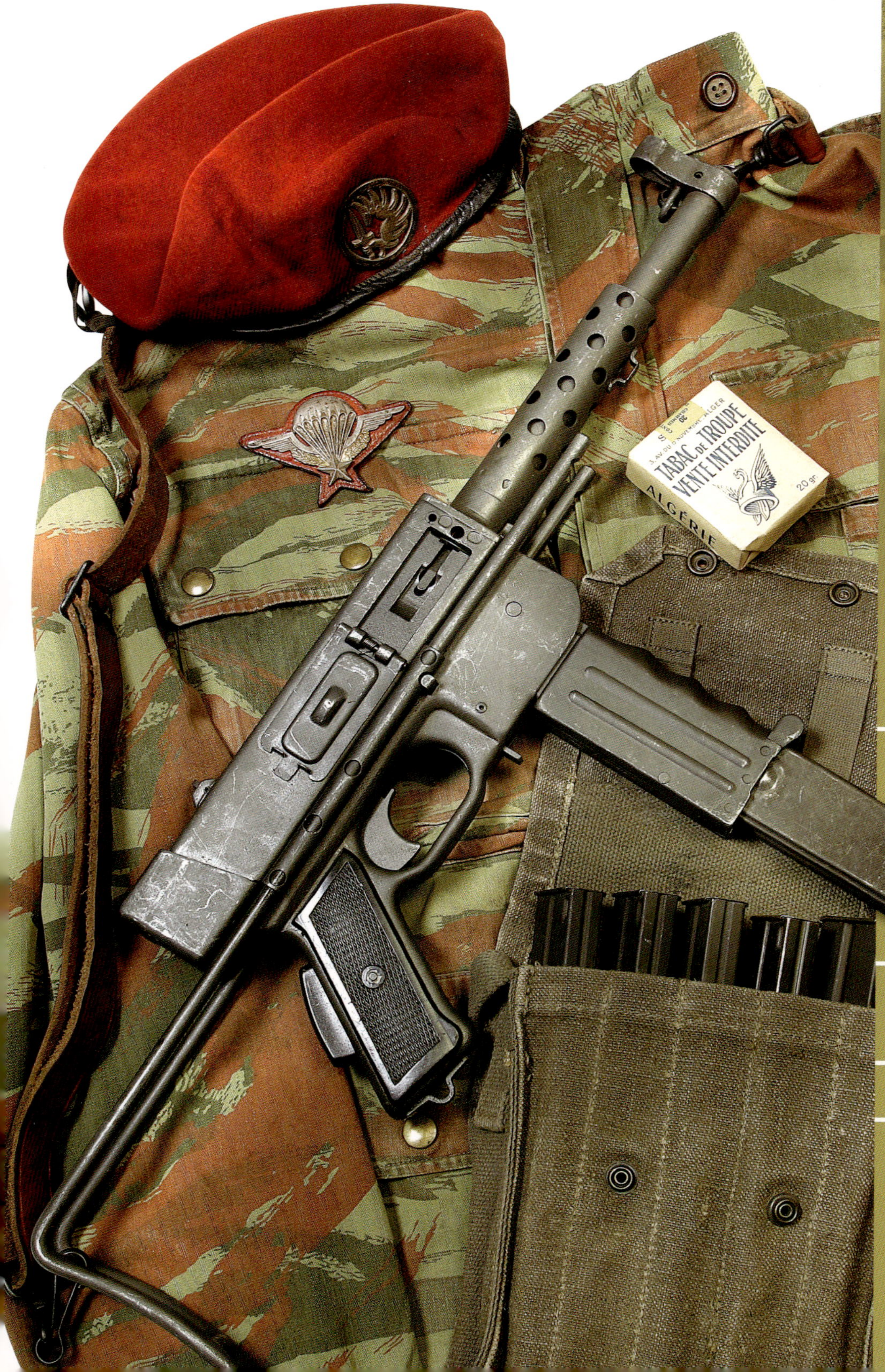

OTHER FRENCH SUBMACHINE GUNS IN 9 MM PARABELLUM

7.65 MM LONG CARTRIDGE

MAS-35 SUBMACHINE GUN

AMMUNITION

ACCESSORIES

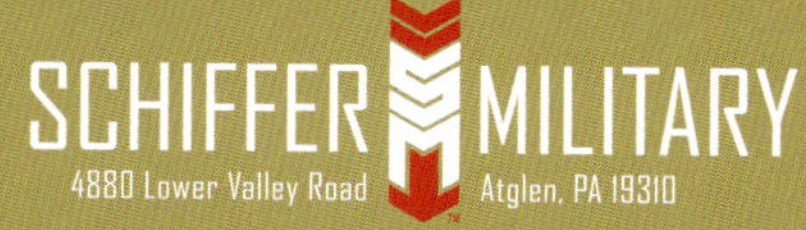
SCHIFFER MILITARY
4880 Lower Valley Road Atglen, PA 19310

Originally published as *Les P.M. français et le MAT 49* by Regi'Arm,
Paris © 2007 Regi'Arm
Translated from the French by Julia and Frédéric Finel

Library of Congress Control Number: 2020952743

Cover design by Justin Watkinson
Type set in Helvetica Neue LT Pro/Times New Roman

ISBN: 978-0-7643-6292-7
Printed in China

Published by Schiffer Publishing, Ltd.
4880 Lower Valley Road
Atglen, PA 19310
Phone: (610) 593-1777; Fax: (610) 593-2002
E-mail: Info@schifferbooks.com
Web: www.schifferbooks.com

CONTENTS

CHAPTER 1

A NEW TYPE OF WEAPON: THE SUBMACHINE GUN

From the 1920s onward, the Bergmann MP 18/I was modified, by the firm Haenel of Suhl, in order to be fed by a straight magazine covered by a Schmeisser patent.

The men of the French "Corps Franc" of Capt. Erhardt in the park of Lustgarten in Berlin during the Kapp putsch of March 1920. An SMG 18/I is attached under one of the heavy load transport carriages. This weapon had proved its worth in trench combat and would soon find a use in street combat. *DR*

During the last months of the First World War, the Germans put into service a weapon that could be considered as the first submachine gun: the Bergmann MP 18/I, a small carbine fed by a drum magazine of thirty-two rounds of P.08 long, capable of firing 9 mm Parabellum on full-automatic (full-auto) setting.

THE LESSONS OF CONFLICT

Other nations were developing similar concepts at the same period:

- The Italians had adopted the Villar Perosa light machine gun in 1915. This double-barrel weapon fired a pistol cartridge: the 9 mm Glisenti, very similar to the German 9 mm Parabellum although less powerful. It rapidly became clear that the two tubes of the Villar Perosa could be uncoupled and each one mounted on a stock to create a light and maneuverable machine pistol: the Beretta 1918 model, also known as the "Revelli machine gun," named after the inventor of the Villar Perosa, Col. A. B. Revelli.
- The Austro-Hungarians had concentrated their efforts in another direction: the transformation of their Steyr 1912 model pistol into a weapon capable of full-auto fire.
- In the United States, Gen. John T. Thompson worked on the development of a .45-caliber (11.43 mm) submachine gun, nicknamed "Trench Broom." The standard version of the 1921 Thompson model submachine gun was not available in time to take part in the first world conflict.

In the second half of World War I, the French army concentrated its technical resources on the development of three new types of infantry weapons: the 1915 light machine gun, the VB rifle grenade, and the FSA 1917 and 1918 semiautomatic rifles. The submachine gun was not a priority for research for either the French or the British during the war.

As soon as the conflict ended, it became clear that in France, a complete overhaul of armaments should be undertaken. Weapons and ammunition adopted before the beginning of the conflict were indeed technically outdated, and models developed during the war had been done so in haste and their manufacture carried out in the context of shortages both of industrial resources and raw materials.

Steyr 1912/16 pistol with fixed thirty-two-round magazine and fire mode selector. *Thomas B. Nelson*

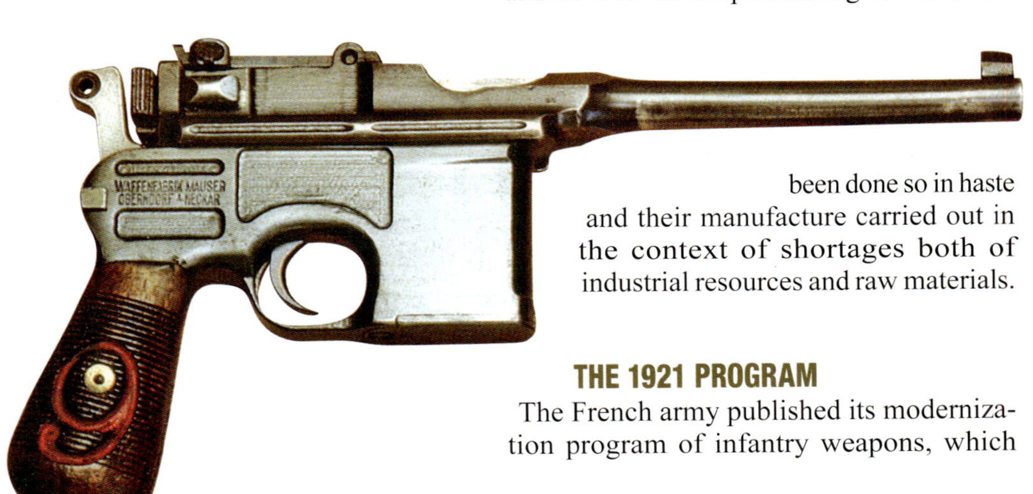

Mauser C.96 pistol

THE 1921 PROGRAM

The French army published its modernization program of infantry weapons, which

The Italian model 1918 submachine gun in Glisenti 9 mm caliber entered into service with the Italian assault troops (Arditi) in the first months of the First World War. This weapon, a direct descendant of the Villar Perosa light submachine gun, was made by the Beretta company. The engineer Tullio Marengoni, who went on to create numerous models of Beretta submachine guns, ensured the development of this model, resulting in a modification of the Villar Perosa light submachine gun, created by Colonel Abiel Betel Revelli. Unlike pistols transformed for full-auto fire, such as the Steyr 1916 model, which has the appearance of a sidearm for close-quarter combat, the Revelli and the German 18/I submachine gun are offensive weapons conceived for attack. As can be seen, these two approaches concerning the use of the submachine gun (as an offensive or defensive weapon) have coexisted since this type of weapon came into being.

Immediately after the First World War, the Oficine di Villar-Perosa produced this submachine gun based on the mechanism of the Villar Perosa light submachine gun 1915 model. The weapon is equipped with two triggers, one of which controls full-auto fire, and the other, single-shot fire.

determined the military characteristics of seven new types of weapons:

- semiautomatic rifle
- automatic pistol
- submachine gun
- heavy machine gun
- antiaircraft machine gun
- portable machine gun
- small-caliber antitank weapon

The technical specifications of the submachine gun were largely based on the examination of captured German 18/1 submachine guns, whose features had been considered of interest with the exception of its magazine. On this point, the slightly curved magazine, containing cartridges in two columns in a double-stack, double-feed arrangement with alternate distribution on the lips of the Italian Villar Perosa submachine gun and later on the Revelli, seemed preferable to the French technical departments.

The 1921 program thus defined the future submachine gun of the French army:

The submachine gun is a light weapon, with light ammunition, able to provide high density fire at short distances (200 meters):

- The weapon was to have the shape of a shortened carbine.
- Its weight was to be between 3 and 4 kg.
- Its ammunition had to be the same as that of the submachine gun chosen as the army model. Until this model was decided on, the weapons to be presented had to fire the 9 mm Parabellum cartridge.
- The weapon would fire magazines of twenty-five rounds at least.

Austrian soldier shouldering a captured Villar Perosa in the style of a submachine gun. *DR*

The Villar Perosa light submachine gun

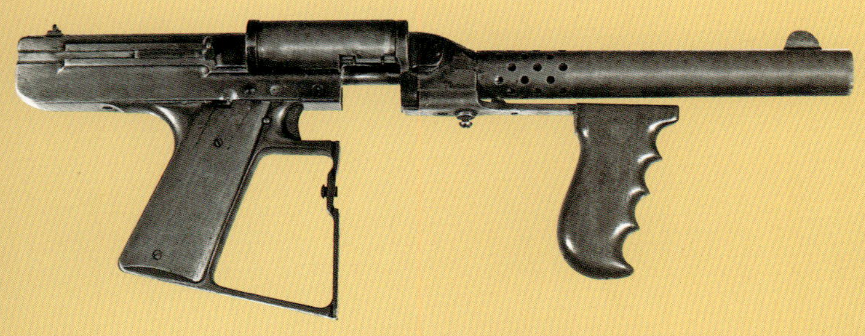

One of the prototypes of the Thompson submachine gun: "the Persuader." This gun in .45 ACP (11.43 mm) caliber, equipped with a delay opening mechanism of the bolt designed by Cmdr. John N. Blish of the US Navy, was developed under the general management of John T. Thompson and was to be used in the trenches. It was christened the "Trench Broom." Its development, however, was too late for it to take part in the First World War. Diverse variants of the Thompson submachine gun were tested by the French army between the wars. *West Point Military Museum*

The majority of magazines used up to the present day in SMGs are derived from the Villar Perosa, in which cartridges are presented alternatively on each lip, or a Schmeisser-type magazine, in which both columns of cartridges interlock into one toward the top of the magazine to present the top cartridge in a central position on the way of the moving bolt. The magazines of French machine pistols conceived before the war were of the Villar Perosa type (which French technical services designated as the "Revelli magazine"). The MAS-38 submachine gun magazine was based on this principle. After the Second World War the French army opted for a Schmeisser-type magazine, with the MAT-49 submachine gun magazine representing a distant descendant.

- Its normal firing mode was fully automatic; a mechanism allowing single-shot firing was not requested.
- Firing speed was to be between 400 and 500 shots per minute.
- The weapon was to be very basic, simple, and well protected against mud.
- The principle of a weapon with moving bolt was recommended, since it was simpler.
- The weapon, fired on a support at a distance of 100 meters, had to give a rectangle smaller than 70/70 (cm) on full-auto fire of five to six rounds, and less than 100/100 (cm) when firing a fully loaded magazine.
- Accuracy in straight-arm shooting would be tested and taken into account in the appreciation of the weapon.
- The rear sight would have 100- and 200-meter notches only
- In order to facilitate firing on a support, the weapon had to be equipped with a bipod. The presence of this bipod should not hinder in any way when engaging in straight-arm firing or with a support.

This very practical document, dated May 11, 1921, was written by soldiers still close to the reality of combat. This book of military characteristics, which will be referred to here under the name "program of 1921," was to guide all French research on submachine guns up to the Second World War.

Throughout the following years, weapons of various origins were proposed to the French army:

- weapons developed and made within national weapons factories, including some prototypes conceived and made with local materials, in various military organizations and establishments, on the initiative of military personnel and technicians who were assigned there
- weapons developed by independent inventors or by private French or foreign enterprises

Two organizations were principally involved in the evaluation of submachine guns up to the Second World War:

Infantryman of the 51st Infantry Regiment armed with a Long P.08. *Maurice Sublet*

Long P.08 pistol

With the appearance of the model 1915 machine rifle (Chauchat), the French infantry was equipped with a light automatic weapon so a soldier could fire while walking. The mobility of the Chauchat, far superior to that of its German equivalent the MG 08/15 light machine gun, is indicative of the fact that the French army was less interested in the development of a submachine gun than was the German army.

From the end of the First World War, the French army dedicated a maximum of effort to the development of a new machine rifle and its ammunition. This research rapidly led to the development of the model 1924 machine rifle in 7.5 mm caliber, which was modified in 1929 to use a slightly shorter cartridge. Reliable, compact, and very mobile, the FM 24-29 gave total satisfaction to all its users. The infantry section of the French army was organized around the service of this weapon. In this context the policy of use of the submachine gun remained relatively unclear in France until the Second World War.

- The CEV (Commission d'Experience de Versailles, or Testing Commission) was within the ETVS (Etablissement Technique de Versailles, or Versailles Technical Establishment), which carried out the first evaluation of light weapons of all origins proposed to the French army. This evaluation was essentially technical in nature:

1. study of the operational principles of the weapon and its manufacture
2. operational tests with measurement of firing accuracy and speed
3. analysis of recorded firing incidents
4. summary of the advantages and flaws of the weapon and a notice concerning its value for the armies, along with its compliance with the directives of the 1921 armaments program and its later modifications

The Versailles Technical Establishment also studied, in the context of what is referred to today as "technology surveillance," various foreign weapons that have never been officially proposed to the French army but were retrieved and confiscated or through various other channels (military attachés, intelligence services, etc.).

The Infantry Testing Commission (CEI) was set up in Mourmelon camp, whose role was to assess the weapon in the field and also possibly to specify its tactical use. In 1919, the CEI was succeeded by the ENT (Ecole Normale de Tir) or Firing School, also set up in the Mourmelon camp (called "Chalons camp").

For weapons developed with French military organizations, the tests were initially carried out on one or two prototypes. When the CEV or CEI considered that the weapon satisfied the demands of the army, it was ordered that a limited series be produced (in general, from fifty to two or three hundred specimens), in order to carry out a wider-based experimentation within the units.

The manufacture of these small series was carried out in state factories:

- the MAC or National Arms Factory of Châtellerault
- The MAS or National Arms Factory of Saint-Etienne. In this particular case, it was a specific section of this establishment, the "test section," whose role was to make prototypes and limited series to be used for tests.
- the MAT or National Arms Factory of Tulle

In the event of a model of weapon being adopted definitively, these factories would ensure the mass production of the weapon or some of its components either alone or in association with other manufacturers.

The Chauchat in action.
DR

CHAPTER 2

FRENCH SUBMACHINE GUNS IN 9 MM PARABELLUM

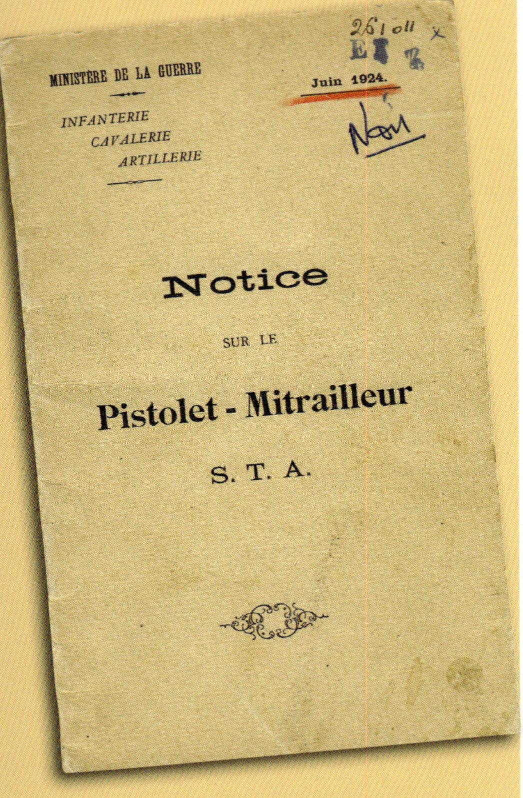
MINISTÈRE DE LA GUERRE

Juin 1924.

INFANTERIE
CAVALERIE
ARTILLERIE

Notice

SUR LE

Pistolet - Mitrailleur

S. T. A.

Provisional notice relating to the STA submachine gun, dated June 1924

STA SUBMACHINE GUN

When the weapons program was published in May 1921, the development of a submachine gun carried out by the Artillery Technical Section (STA) was already virtually complete, as testified by a letter from July 12, 1921, signed by General Mochot, technical research and development inspector of the artillery, indicating that "the experimentation, undertaken by the research commission of Versailles on the specimen of submachine gun established by the technical section of the artillery, having given satisfactory results, it seems necessary to allow more complete research to build other specimens of this weapon."

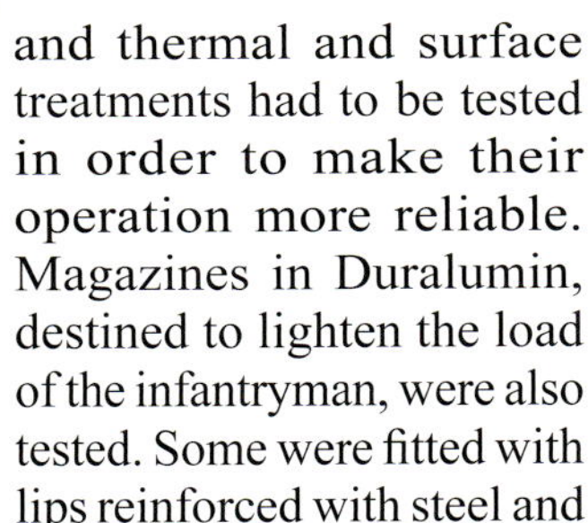
Bolt assembling the rear of the receiver in the frame of the STA submachine gun

Although the development of the weapon was relatively rapid, the development of the magazines caused enormous difficulties. The magazines were made by the National Weapons Factory at Châtellerault (MAC). Between 1921 and 1926, multiple types of sheet metal and thermal and surface treatments had to be tested in order to make their operation more reliable. Magazines in Duralumin, destined to lighten the load of the infantryman, were also tested. Some were fitted with lips reinforced with steel and only curved, and thirty-two-round magazines made completely of steel were selected due to their shock-resistant capacities.

In 1924, the general staff published a research program in combat units and placed an order with the weapons factory at Saint-Etienne (MAS) for 300 STA submachine guns intended to be split between the infantry units (150), artillery (80), cavalry (40), combat tanks (10), and in research commissions (10). The remaining ten weapons were to be kept in reserve at the weapons factory at Saint-Etienne.

The STA submachine gun proved to be a reliable, accurate weapon that was also easy to disassemble and maintain. However, some modifications to improve it were proposed:

- Replace the small-diameter recoil spring, which had a tendency to become misaligned, with a larger-diameter spring.
- Replace the upward-facing, hook-shaped bolt handle with a cylindrical or square knob.
- Modify the bolt cover.

In this modified version, called M1, the STA submachine gun would doubtless have been a remarkable combat weapon and would have allowed the French army, from 1925 onward, to have at its disposal a weapon broadly equivalent to those used during the Second World War and even later!

A flap allowing the bolt handle to be blocked forward on an STA. This flap both improved weapon safety during transport and prevented foreign bodies from entering the mechanism.

The German MP 18/I submachine gun and the French STA submachine gun, the latter of which takes up a certain number of features of the mechanism of the German weapon, while its curved, double-column magazine was roughly based on the Italian Revelli. A protective cover (*Schutzkappe*) for the lips of the thirty-two-round drum magazine is to the left of the 18/1. To the left of the STA is a spare magazine and a box of German 9 mm Parabellum ammunition. These were in fact German cartridges, captured in 1918 and used for testing the French SMGs in this caliber, and were even used in service in 1940 to load the EMP and other SMGs in 9 mm caliber used at that time in the French "Corps Francs." ***Marc de Fromont***

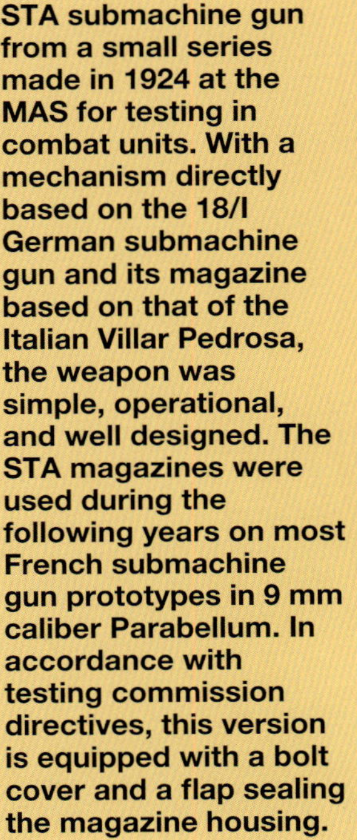

STA submachine gun from a small series made in 1924 at the MAS for testing in combat units. With a mechanism directly based on the 18/I German submachine gun and its magazine based on that of the Italian Villar Pedrosa, the weapon was simple, operational, and well designed. The STA magazines were used during the following years on most French submachine gun prototypes in 9 mm caliber Parabellum. In accordance with testing commission directives, this version is equipped with a bolt cover and a flap sealing the magazine housing.

Drawing from a provisional user notice showing an STA machine pistol from the series destined for tests in combat units, with a tubular bipod

However, only a few specimens of the STA were modified to the M1 standards for testing. The appearance of more-compact submachine guns such as the MAS-1924, but above all the choice made in 1926 by the French army to adopt the 7.65 mm Long cartridge ammunition common to future pistols and submachine guns of the French army, put an end to the career of the STA and STA M1 submachine guns, the manufacture of which was discontinued.

During the following fifteen years after the publication of the 1921 weapons program, the SMG, initially considered as an offensive weapon destined for close-quarter combat, was gradually reduced to the role of a self-defense weapon, destined to arm noncommissioned officers (NCOs) and officers, artillerymen (for the protection of their field guns), armored-vehicle and machine gun crews, and various other specialists whose principal purpose did not require the use of an individual weapon.

This resulted in the final choice of the 7.65 mm Long over the 9 mm Parabellum as the common ammunition for both the pistol and the submachine gun, so as to reduce the load of transporting ammunition for the users of the latter.

Exploded-view diagram of the bolt

Prototype of an unidentified submachine gun, made from an STA. *Jacques Barlerin*

Right-side view of the CEI No. 1 submachine gun. *Jacques Barlerin*

Second type of CEI submachine gun

INFANTRY TESTING COMMISSION SUBMACHINE GUNS

These two prototypes, called no. 1 and no. 2, were developed with the Infantry Testing Commission (CEI), probably under the initiative of military personnel and technicians assigned to this organization, which tested them in 1925.

The no. 1 had a rate of fire that was too fast (around 1,000 shots per minute). No. 2 was fitted with a heavier bolt mechanism, which had a longer movement. Even though this arrangement reduced the firing speed to around 650 shots per minute, it led to irregular functioning and firing incidents. After the Versailles Technical Establishment recommended against additional research in this area, the Infantry Testing Commission did not propose any new prototypes.

MAS 9 MM CALIBER SUBMACHINE GUN

The MAS (National Arms Factory at Saint Etienne), which had already provided the manufacture of the STA submachine guns, did not delay in proposing its own models to the Versailles test commission. One of the objectives pursued by the MAS was to develop a shorter weapon than the STA submachine gun, but without reducing the length of the barrel.

The reduction of the length of the receiver translated into a reduced movement of the bolt, bringing with it the risk of increasing the rate of fire. Two avenues of research were undertaken by the MAS:

1. To compensate for the reduction of the length of the receiver by housing the recoil spring in the stock of the weapon; this principle was included in prototypes no. 1 and no. 1a. This disposition, which initially seemed astute, ultimately proved to be a disadvantage when the development of motorized and airborne units brought to light the value of having a weapon with a folding stock. This development was almost fatal for the development of the MAS submachine gun.

2. To use a relatively short recoil spring housed in the receiver, but to insert a device to reduce the rate of fire. This was included in prototypes no. 2 and no. 3. In this configuration, the fact

This officer is firing a MAS submachine gun in 9 mm, recognizable by its curved magazine.

Top: SE No. 1 MAS 1924 submachine gun. **Bottom:** SE No. 1a MAS 1924, both in 9 mm caliber Parabellum. No. 1 is fitted with a barrel without cooling vent. The barrel of the no. 1a is, however, protected by a perforated cooling vent. *Jacques Barlerin*

The no. 1 had the interesting particularity of a bolt with a wide notch, in which the user placed a finger to bring the bolt back during the arming of the weapon. This feature was later taken up on the American M3A1 submachine gun. The no. 1a (*bottom*) had a sealing flap on the ejection port. The arming of the bolt was ensured by a standard bolt handle positioned on the right side of the weapon. This lever was solid, with an obturating flap that prevented foreign bodies from entering the slide mechanism groove—an arrangement that would be found on the majority of French submachine guns. *Jacques Barlerin*

of not housing a recoil spring in the stock meant that the adoption of a weapon with a folding stock could be envisaged, an avenue that was explored with the no. 2 prototype.

The MAS submachine guns presented to the Versailles Technical Establishment in 1924 were chambered for the 9 mm Parabellum cartridge. They were fed by the same curved magazine as the STA submachine guns. The bolt on the first models was also very close to that on the STA.

These weapons operated with a moving bolt, firing with an open bolt. They were equipped with a grip with two triggers (controlling the choice of single-shot or full-auto fire). This grip closely resembled the one on the 1924 FM model, which was starting to come off the production lines of the national weapons factory of Châtellerault (MAC).

The 6-degree angle between the axis of the barrel and that of the bolt recoil engendered a braking action in the latter, which allowed a reasonable rate of fire to be conserved.

These weapons were made by the "test section" (abbreviated to SE) of the MAS. This section had both the equipment and the personnel to make small series, up to a few hundred specimens, of the prototypes used for tests. It was distinct from the "manufacturing" section, whose role was the large-scale production of models adopted by the army.

The prototypes conceived by the MAS and made by the "test" section are identifiable by the addition of the abbreviation SE to their name.

MAS-1924 SE MODEL NOS. 1 AND 1A

The MAS-1924 SE model Nos. 1 and 1a submachine guns were more compact than the STA submachine gun. During the tests they proved to be more stable in full-auto fire mode. However, their operational safety never reached the level of the STA, and consequently they were not retained either by the Versailles Technical Establishment or by the Infantry Testing Commission due to their erratic operation.

MAS-1924 SE MODEL NOS. 2 AND 3

These two models were reduced versions of the STA submachine gun, fitted with a firing-rate-reducer system to avoid any increase in the rate of fire despite a reduction in length of 14 cm. The no. 2 is fitted with a metal folding stock, while the no. 3 has a standard wooden one.

The back-and-forth movement of the bolt was not transmitted to the cocking handle.

This design meant the shooter had to push the bolt handle back in order to engage it with the bolt, before being able to arm the submachine gun.

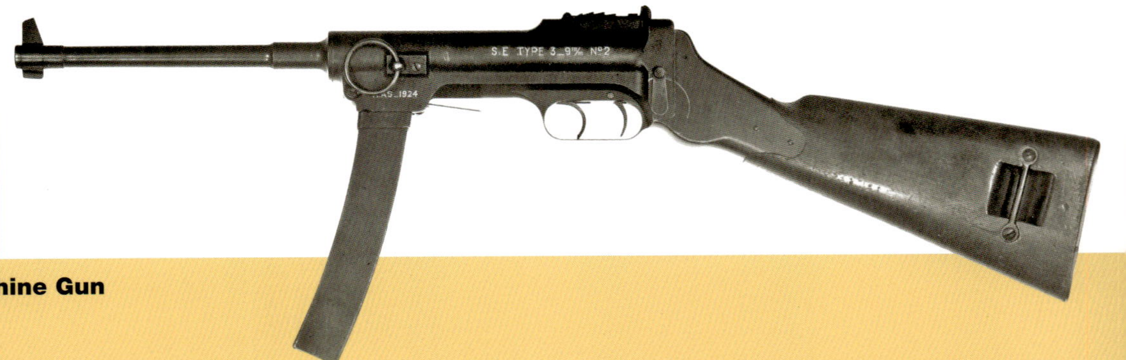

MAS-1924 SE No. 3 submachine gun. This standard-design weapon was derived from the STA. In order to make a shorter weapon than the STA, the MAS fitted this model with a firing-rate reducer. *Jacques Barlerin*

MAS-1925 SE submachine gun: a very successful weapon, which unfortunately was not adopted due to the abandonment of the 9 mm caliber Parabellum in favor of the 7.65 mm Long by the French army. Like all other MAS submachine prototypes of this period, the weapon used an STA magazine. *Jacques Barlerin*

This device was taken up on the MAS-1925 model SE submachine gun.

Experiences of the weapon falling on the stock showed that firing took place as soon as the bolt was moved back sufficiently for a cartridge to move into the chamber.

MAS-1925 MODEL SE SUBMACHINE GUN

This is an improved version of the MAS-1924 model SE No. 1, taking into account the modifications requested by the Infantry Testing Commission that resulted from the tests on the model 1924 SE No. 1 and No. 1a. The weapon is both lighter and shorter than previous models. The single-shot function had been removed, as had the perforated protective barrel jacket. The cocking handle is the MAS SE No. 3 type. During use, the fact that the shooter had to push the bolt handle forward to lock it in the bolt before arming his submachine gun was justifiably considered by the CEI to be a source of hesitation and loss of time for the shooter.

The MAS-1925 model SE presented a reliability in its operation that was far superior to those of previous MAS prototypes. During tests, only 0.7% of operational incidents were recorded, and the majority of those incidents were perhaps due to the deplorable condition of the captured German ammunition used for the tests. The test reports of the model 1925 SE submachine gun highlight the fact that cartridges were often oxidized and that 20% of the cases were split lengthwise (a remark noted in the majority of test reports of the French 9 mm submachine guns of the period).

As will be discussed later, when the model 1925 tests took place, the French army had already chosen to chamber its future automatic pistol and submachine gun in 7.65 mm Long caliber and not in 9 mm Parabellum. Experiments continued for several more years with prototypes in 9 mm caliber before the manufacture of this cartridge began in France.

MAS-1932 MODEL 9 MM CALIBER SUBMACHINE GUN

A small order of fifty weapons in 9 mm caliber Parabellum of a very slightly modified version of the MAS-1925 model SE submachine gun was made in 1932 under the name "MAS-1932 model submachine gun in 9 mm."

This version was different from the initial 1925 model by the presence of an eyesight in the form of a bracket gauged for distances of 100 and 200 meters and a double trigger grip.

The weapon was made on the request of the Ministry of the Colonies to be put into service overseas (in China and French West Africa).

Even though at that time the French army had already chosen the 7.65 mm Long as ammunition for its future submachine gun, the Ministry of the Colonies preferred to order a weapon chambered in 9 mm caliber Parabellum, whose cartridge had been finalized for a long time, and with which the MAS submachine gun presented a satisfactory level of operational safety (provided that cartridges in good condition were used!).

The model 1932 submachine gun could be worn in a case at the belt. The pouch for accessories and spare parts was worn on the belt carrying strap; the pouch for magazines was attached to the belt. *DR*

1932 SE model MAS submachine gun. This weapon, also in 9 mm Parabellum, was very close to the 1925 SE MAS. Fifty were made for the Ministry of the Colonies. Unlike that of the 1925 SE MAS, its sight had just one bracket with two eyepieces. *Jacques Barlerin*

CHAPTER 3

THE FRENCH ARMY ADOPTS THE 7.65 MM LONG CARTRIDGE

1924 type 1 MAS (*top*) and type 1a (*bottom*) in long 7.65 mm caliber. From the outside the type 1 is identical to type 1 in 9 mm Parabellum. Type 1a in 7.65 mm, however, has a long barrel protected by a jacket. Measurements of initial speed and accuracy would show that the long barrel had no advantage over the 215 mm barrels used for other models, and this led to it being abandoned. Note also the simplification of the rear sights compared to 9 mm caliber models. *Jacques Barlerin*

From the end of the Great War, the French army had envisaged abandoning service handgun cartridges in favor of a new, more powerful cartridge that would be common both to the future submachine gun and the future automatic pistol, the adoption of which was contained in the 1921 armament program. In anticipation of a definitive choice of this ammunition, it had been decided that the prototypes of the submachine gun would be chambered in 9 mm Parabellum. Very large numbers of cartridges of this type captured from German troops were available in French arsenals. Private inventors wanted to present a submachine gun to the French army that could be assigned cartridges of this caliber by the War Ministry in order to develop their prototypes.

There was, however, no particular reason for France to choose to definitively adopt the 9 mm Parabellum, which at that time was made only in Germany, rather than any other cartridge. Although the 9 mm Parabellum became the principal NATO submachine gun and pistol cartridge just after the Second World War, it was just one ammunition among many others between the wars.

A project to begin manufacturing the 9 mm Parabellum was researched at the Valence cartridge factory, but the general staff finally opted for a 7.65 mm caliber with long case (called 7.65 mm Long), identical to the American .30 Pedersen ammunition, in 1925.

This little-known cartridge had been designed to feed a device allowing for the regulation American Springfield 1903 rifle to be transformed into an automatic weapon. It was more powerful than the 7.65 mm Browning, commonly used during the Great War by the French army in the Ruby, Star, and Savage pistols bought as a complement to the model 1892 revolver.

The technical services of the army had tested the .30 Pedersen cartridge in 1922 by studying a semiautomatic carbine, in this caliber, that John M. Browning had proposed to the French army.

The choice of the 7.65 mm Long ammunition was frequently criticized in contemporaneous publications. This decision relied, however, on arguments that were pertinent at the time, summarized in this extract from an ETVS report:

FROM LEFT TO RIGHT: .30-caliber Pedersen Long cartridge with brass case and jacket in cupronickel (marking RA 19); two short cartridges in .30-caliber Pedersen with brass case and jacket in tombac (RA H19 and RA 19). *Loïc L'Helguen*

FROM LEFT TO RIGHT: **German 9 mm Parabellum cartridge, 7.65 mm Browning cartridge, 7.65 mm Long (this example was made in 1952), and an 8 mm cartridge for an 1892 model revolver**

This choice was justified by the almost identical performance to the 9 mm Parabellum in terms of accuracy and the perforation up to 600 meters for a lighter weight of the 7.65 mm Long (9.3 g compared to 12.2 g for the 9 mm Parabellum cartridge). This difference in weight meant the supply of 7.65 mm cartridges could be increased by one-third compared to those in 9 mm for a specific weight.

In addition, due to the lesser reaction it transmitted to the weapon, the 7.65 mm cartridge meant the weight of the pistol was reduced by between 200 to 300 grams.

Another advantage of this ammunition: it could be rapidly put into production in the United States, whereas the 9 mm Parabellum was not produced either in France or the United States at that time. The experience of the First World War had convinced the French army of the necessity of being able to appeal to the power of production of American industry to finish its own productions.

Fifty thousand .30 Pedersen cartridges were ordered from the United States to carry out the initial tests. In order to distinguish it from the 7.65 mm Browning, commonly used in France for Ruby- and Star-type regulation pistols, this cartridge was called "7.65 mm Long."

Production in France did not start before 1930 by the French Munitions Company (SFM).

SE MAS 1935 submachine gun bearing the serial number 5. The locking bolt, which tended to loosen up and get lost, would be abandoned on the war production MAS-35 SE. ***Private collection***

A DIFFICULT DEVELOPMENT: THE MAS SUBMACHINE GUN IN 7.65 MM LONG

With the adoption of the model 1924 light machine gun, which five years later was to become the 1924-M29 model, the combat group found itself organized around a particularly high-performance and effective weapon. The submachine gun from that point on began to lose its importance as a support weapon, especially since the projects for adoption of semiautomatic rifles, of the same caliber as the light machine gun, were already opening the way for a considerable increase in firepower of the combat group.

In the difficult financial context of the interwar period, the general staff preferred to concentrate resources on the manufacture of the FM 24-29 and on the development of modern individual shoulder weapons, firing the same ammunition as the FM: a manual

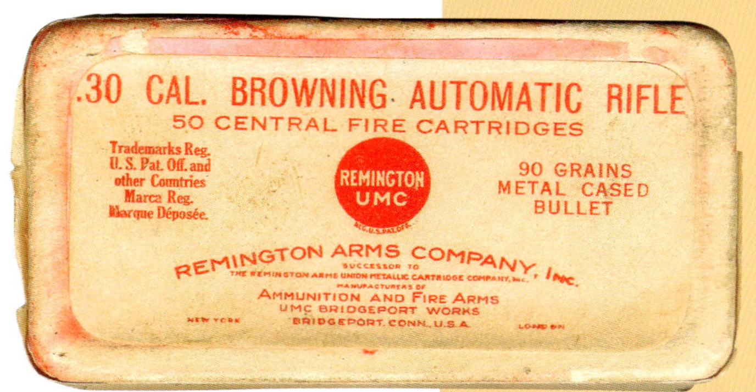

RIGHT: Box of 7.65 mm Long, part of the first production carried out in France by the SFM. LEFT: Box of cartridges for the Browning automatic rifle .30 caliber. ***Jacques Barlerin***

MAS-24 submachine gun, 7.65 mm caliber

SE MAS-1931 model submachine gun in 7.65 mm Long. This is a version in 7.65 mm long of the SE MAS-1925 model with a controlled firing pin. *Jacques Barlerin*

repeat-firing rifle that led to the MAS-36, and a semiautomatic rifle that led to the MAS-40, with the French defeat in 1940 preventing its mass production.

In these conditions, the submachine gun came to be increasingly consigned to the role of a self-defense weapon, for which the use of an individual weapon was not the primary function: for vehicle crews, gendarmes, and all the specialists: artillerymen, signalers, et al. Therefore, it was important that it was as light and easy as possible to use: the choice of 7.65 mm Long caliber was part of this process. It was completed in the mid-1930s with the requirement of having a folding stock and magazine.

As a self-defense weapon it became one of second necessity. The decision was made to classify it with those weapons whose manufacture should commence only after mobilization.

This choice, coupled with the limited means dedicated to the development of this type of weapon, delayed the adoption of a definitive model up to the Second World War.

Between 1924 and 1931, the MAS presented some versions of no. 1 and no. 1a and the 1925 model in 7.65 mm Long to the ETVS and the CEI. These weapons, which functioned very well in their 9 mm Parabellum versions, lacked operational safety in the 7.65 mm Long version.

During this 1924 to 1939 period, many submachine gun prototypes were devised by the research section of the MAS, but the majority of them were not elaborated on or developed, and some of them were never even presented to the ETVS for evaluation. Some photos are shown, found by Jacques Barlerin in the archives of the MAS, so these prototypes are not completely forgotten.

In 1935, the MAS proposed a new submachine gun in 7.65 mm Long, known under the name "SE MAS-35," to be examined by the ETVS and the CEI. This weapon was based on the 1924 and 1925 models that had already been tested in this caliber. Compared to the previous models, the SE MAS-35 had many improvements. It had a reduced-diameter bolt (22 mm instead of 30 mm) and a firing pin, independent of the bolt, fixed by a pin.

SE MAS-1926 model no. 0 in 9 mm caliber. Interestingly, while all the other MAS submachine guns in 9 mm caliber are fitted with a curved STA magazine, this specimen is fitted with a straight magazine. *Jacques Barlerin*

The bolt recoil axis was now at an angle of only 5 degrees with that of the barrel, and not 6 degrees as on previous models. The tilt of the magazine in relation to the axis of the barrel was reduced to 9 degrees (against 12 degrees on previous models). The rear sight was from then on composed of two independent eyepieces disappearing completely in the receiver for transport purposes.

As a result of the tests carried out in 1936 at the Mourmelon camp, the CEI called for the correction of several details and recommended that a small series of SE MAS-1935 submachine guns be made, with the aim of testing them in combat units. In application of this recommendation, the DEFA (Weapons Research and Production Unit), which had recently been created to coordinate the activities of the weapons industry in France, placed an order with MAS by letter on November 13, 1937, for 500 magazines along with the accessories and spare parts considered useful for testing in combat units.

Despite this first step toward an official adoption, the MAS-1935 model was faced with two formidable competitors: the ETVS and the Petter. These two weapons, of a more recent conception, had the advantage over the MAS submachine gun of having a folding magazine well under the barrel and a fold-back stock. As will be seen, only the outbreak of the Second World War prevented the MAS submachine gun from being definitively abandoned in favor of the Petter submachine gun.

Remains of a MAS-31 submachine gun tested with a bolt made in two parts, with opening delay. *Henri Canaple*

Features of the Principal Prototypes of French Submachine Guns
In 9 mm caliber Parabellum, made by the MAS from 1924 to 1938

Model	STA	MAS type 1	MAS type 1a	MAS type 3	MAS-1925	MAS-1932
Total length	855	640	640	710	630	580
Barrel length	210	225	225	225	220	180
Mass	3.4	2.97	3.35	2.7	2.7	2.8
Rate of fire	380	450	480	?	550	500

Features of the Principal Prototypes of French Submachine Guns
In 7.65 mm caliber Long, made by the MAS from 1924 to 1938

Model	MAS type 1	MAS type 1a	MAS-1925	MAS-1935	MAS-1938	MAS-1932
Total length	640	750	630	630	630	580
Barrel length	225	340	220	220	220	180
Mass	2.7	3.2	2.7	2.9	2.9	2.8
Rate of fire	470	500	550	640	640	500

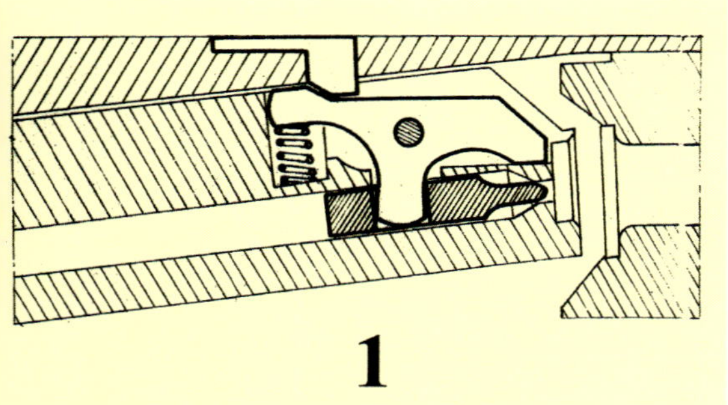

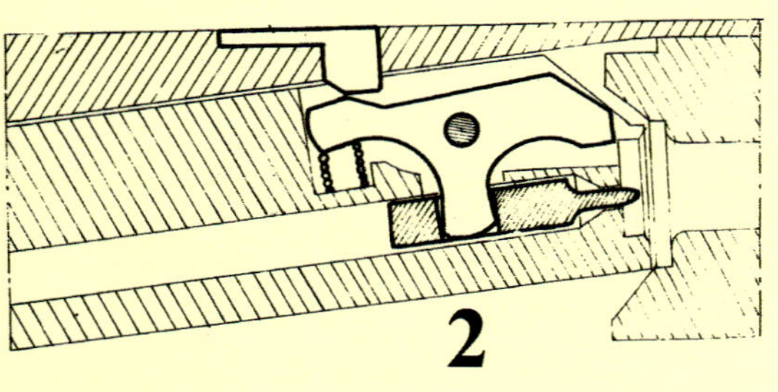

Comparison between the first model of MAS-34 bolt, with firing pin pinned to the bolt (*top*) and the modified model, equipped with a cam bringing about firing when the bolt was completely closed. *Jacques Barlerin*

SE MAS-34 submachine gun. *Jacques Barlerin*

TOP: MAS-36 submachine gun in 7.65 mm Long. This weapon had a bolt in three parts (the bolt itself, pivoting block, and reducing wedge) to ensure a delay in opening. It was presented to the ETVS on December 9 by the director of the MAS to delegations of various sections and interested ministry departments, without success. *Jacques Barlerin*
MIDDLE: SE MAS-36 submachine gun bearing the serial number 2. The chief engineer, director of the MAS, presented this prototype in person to the CEI on December 9, 1936. The weapon performed correctly apart from several nonfirings. It was returned to the MAS after firing 320 cartridges to perfect its development, but this avenue of research was ultimately not pursued. *Jacques Barlerin*
BOTTOM: Another prototype in 7.65 mm Long, bearing the marking SE MAS-38 number from series 1. The weapon is fitted with a cylindrical barrel jacket, nonperforated, the use of which remains unknown. *Jacques Barlerin*

MAS FROM 1920 TO 1940

Numbering of 1935 and 1938 models of submachine guns

Here are the indications given by Daniel Mastier and Stéphane Ferrard in their article published in the *Gazette des Armes* 66:

- SE MAS-35: preproduction of 100 specimens
- SE MAS-35: number followed by the letter F; first production of 1940
- SE MAS-38: number lower than 10,000 F
- Second production of 1940
- SE MAS-38: number higher than 10,000, followed by the letter F—production from 1942 to August 1947
- SE MAS-38: number followed by the letter G; production from September 1947 to August 1949
- SE MAS-38: number followed by the letter H; production from September to December 1949
- SE MAS 38: number followed by the letter FH; last production from 1950

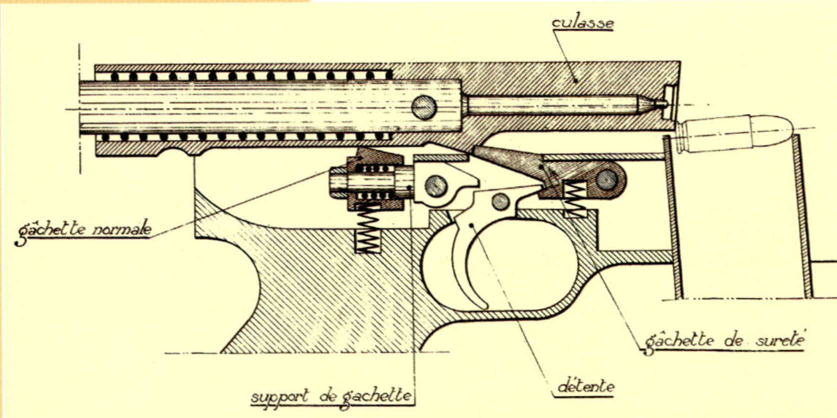

Trigger mechanism, called "safety sear," tested to improve the safety of the SE MAS-35 by preventing the bolt from closing if the trigger was not pressed. This device specifically stopped accidental firing if the weapon was dropped on the stock, leading to the accidental recoil of the moving bolt. *Jacques Barlerin*

CHAPTER 4

A NEW GENERATION OF SUBMACHINE GUNS

The modernization of the French army undertaken in the mid-1930s had to bring about an increased motorization of the infantry, coupled with the development of armored and airborne units. For these troops, the choice of a compact and transportable submachine gun was necessary.

In 1933, the criteria of selection contained in the 1921 weapons program were completed by the addition of two new demands: the submachine guns had to have a folding stock and a folding magazine well. The SE MAS-1935 submachine gun did not satisfy these criteria, although two new, recently developed weapons presented the following characteristics:

- the ETVS submachine gun, from the CEV
- the Petter submachine gun

A certain number of preproduction specimens had to be retrieved by the German occupier since the notices on foreign weapons (*Kennblätter fremden Geräts*) listed the ETVS submachine gun under the name "Maschinenpistole 721(f)."

The ETVS submachine gun, fitted with a folding stock and magazine well, could be transported easily.

THE CEV AND THE ETVS SUBMACHINE GUNS

The CEV

A commission for experiments (CEV) operated within the Versailles technical establishment (ETVS), which also developed its own submachine gun; this was not selected by the CEI during the series of tests carried out in 1936 due to its very high rate of fire, which the pneumatic buffer proved unable to reduce, and therefore the stability of the weapon was compromised during sustained fire.

The CEV submachine gun in 7.65 mm Long caliber; firing with a moving bolt and controlled firing pin unlike the MAS submachine gun, the recoil of the bolt took place in the barrel axis. A pneumatic buffer, housed in the bolt cap, was meant to decrease the rate of fire.

By operating a lever, the CEV could rapidly be disassembled into two parts with a maximum length of 45 cm for transport.

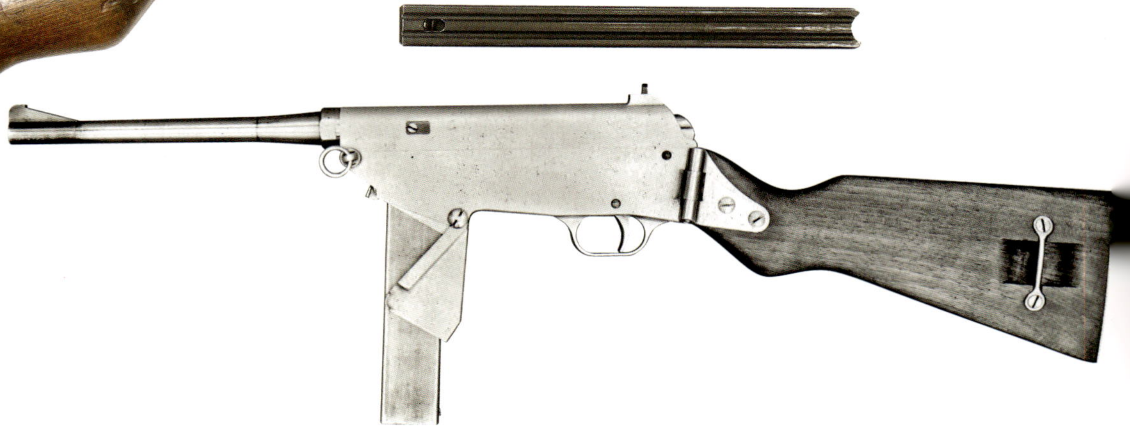

One of the two "nonfunctional" prototypes of the ETVS submachine gun designed by chief engineer Martin

ETVS submachine gun

The Versailles technical establishment rapidly developed a new submachine gun fitted with a folding stock and magazine. Two prototypes of this weapon were built within the establishment.

They were tested in 1937, the first by the ETVS itself and the second by the CEI, along with the Petter submachine gun (presented in the following chapter) and the MAS model 1935 SE submachine gun. Due to its lack of stock and folding magazine, this weapon did not respect the technical specifications and from then on participated in these tests only as a weapon of comparison.

The National Weapons Manufacturer at Châtellerault (MAC) was tasked by the ETVS to make ten preproduction specimens. The barrels, as well as the slightly modified MAS-35-type magazines destined for the ETVS submachine guns, were supplied by the National Weapons Manufacturer of Saint Etienne (MAS).

Because the results of these tests were promising, the DEFA placed an order with the MAC on March 14, 1937, for forty additional ETVS submachine guns. This order was not completed until 1939. Apart from the tests undertaken by the ETVS and the CEI, it had been planned to have some preproduction weapons tested by the "Ministry of Air" special units: in other words the first parachute units in the process of being formed under the aegis of the air force.

By a ministerial decision taken on June 30, 1938, the MAS was tasked with making 600 ETVS submachine gun magazines and fifty-eight barrels to be made at Saint Etienne for the preproduction ETVS series. These magazines were to be made with the machine tools used by the MAS to produce the MAS-35 magazines.

Marking on the ETVS submachine gun

An ETVS submachine gun in a setting that evokes the French army on the eve of the Second World War.
Marc de Fromont

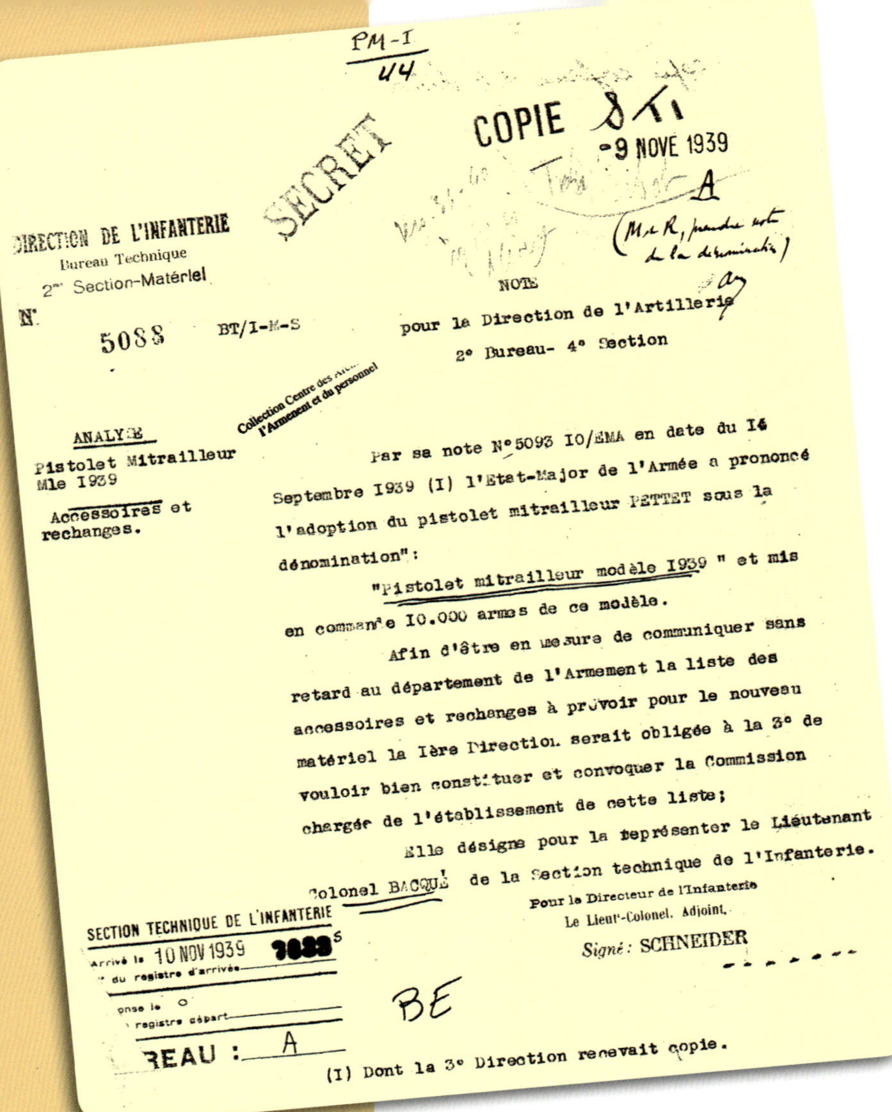

PM-I
44

SECRET

COPIE

-9 NOVE 1939

DIRECTION DE L'INFANTERIE
Bureau Technique
2me Section-Matériel

N° 5088 BT/I-M-S

NOTE
pour la Direction de l'Artillerie
2e Bureau- 4e Section

ANALYSE
Pistolet Mitrailleur Mle 1939
Accessoires et rechanges.

Collection Centre des Archives de l'Armement et du personnel

Par sa note N°5093 IO/EMA en date du 14 Septembre 1939 (I) l'Etat-Major de l'Armée a prononcé l'adoption du pistolet mitrailleur PETTET sous la dénomination": "Pistolet mitrailleur modèle 1939 " et mis en commande 10.000 armes de ce modèle.

Afin d'être en mesure de communiquer sans retard au département de l'Armement la liste des accessoires et rechanges à prévoir pour le nouveau matériel la Ière Direction serait obligée à la 3e de vouloir bien constituer et convoquer la Commission chargée de l'établissement de cette liste;

Elle désigne pour la représenter le Lieutenant Colonel BACQUÉ de la Section technique de l'Infanterie.

Pour le Directeur de l'Infanterie
Le Lieut-Colonel, Adjoint,
Signé: SCHNEIDER

SECTION TECHNIQUE DE L'INFANTERIE
Arrivé le 10 NOV 1939 1883
du registre d'arrivée
BE
BUREAU : A

(I) Dont la 3e Direction recevait copie.

Note from the general staff dated November 9, 1939, mentioning the adoption of the Petter submachine gun under the name of "Model 1939 Submachine Gun," accompanied by the order for 10,000 weapons

PETTER SUBMACHINE GUN

Charles Petter was in fact a Swiss national when he served as a soldier in the Foreign Legion and had climbed the military hierarchy to reach the rank of captain. He had already designed a 7.65 mm caliber Long automatic pistol that was adopted by the French army in 1935 under the name of PA Model 1935 A. This weapon had to be made on behalf of the army by the Société Alsacienne de Constructions Mécaniques (SACM) in its factory at Cholet.

A tireless inventor, Charles Petter also developed an original design for a submachine gun with a stock and magazine that could be folded for transport purposes. The Petter submachine gun had a very short receiver, meaning the bolt could recoil only over a short length. In order to maintain the firing speed at a reasonable rate, the Petter submachine gun was fitted with a decelerator, operated by a spiral spring positioned at the top of the pistol grip. This device ensured a braking of the movement of the bolt during firing, reducing the firing rate to around 570 shots per minute.

The weapon operated with a fixed bolt. Unlocking was carried out by a short recoil of the barrel, followed by the connecting rod lowering the barrel. The system seen here is the locking system of the Colt M1911, also used on the Petter model 35 A weapon.

Throughout the various trials carried out in 1936 and 1937, the Petter submachine gun revealed itself to be less resistant than the ETVS and the SE MAS-1935 when exposed to mud

The adoption of the Petter submachine gun in 1939 and the choice to start production of an improved version of the MAS model 1935 put an end to the career of this very interesting and well-designed weapon.

First model of the Petter submachine gun, recognizable by the cooling holes in the barrel jacket

Second model of the Petter submachine gun, without cooling holes

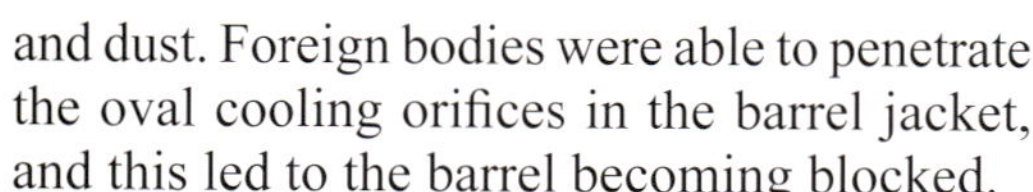

Weapon with magazine folded

and dust. Foreign bodies were able to penetrate the oval cooling orifices in the barrel jacket, and this led to the barrel becoming blocked.

Charles Petter was committed to perfecting his weapon in relation to the results of the successive trials carried out by the ETVS. He added a new safety device based on the one that MAS had developed to improve its model 1935 submachine gun: the trigger could be folded toward the front in a safety position, thereby immobilizing the bolt in a forward position and preventing its accidental recoil, which could cause accidents if the weapon fell onto its stock. He also fitted his submachine gun with a folding stock, strengthened the frame, and removed the catastrophic cooling orifices, instead opting for a barrel jacket that was entirely closed. Modified in this way, the Petter submachine gun had a better handling capability than the MAS 35 and was lighter, meaning that a combatant could transport 50% more cartridges with equal load.

The CEI proposed just one preproduction series of fifty samples, integrating various detailed modifications requested by the CEI and ETVS and ordered from the SACM for trials in combat units concurrently with the MAS-35 and ETVS submachine guns. In anticipation of the manufacture of these fifty submachine guns, the available prototypes were modified in compliance with the testing commission's requirements and used at the La Courtine camp during the summer of 1939. These trials confirmed the qualities of the Petter submachine gun, which was officially adopted under the name "Model 1939 Submachine Gun."

But due to war being declared, the final tests in combat units of the submachine guns in competition (Petter, ETVS, MAS-35) could not be completed, and the SACM, fully absorbed by the production of the model 35A pistols, never made the fifty preproduction models requested by the CEI.

When the stock and the magazine were folded, the Petter submachine gun was extremely compact.

When the stock and the magazine were folded, the Petter submachine gun was extremely compact.

Schmeisser submachine gun tested by ETVS in early 1933, and the Bergmann MP 18/I tested as a comparison. *CAA*

CHAPTER 5

EVALUATION OF FOREIGN SUBMACHINE GUNS BETWEEN 1918 AND 1940

While leading trials of French submachine prototypes, the ETVS made efforts to test the available foreign submachine guns. The provenance of these weapons seems to have been extremely varied. Some were simply presented by manufacturers hoping to see them adopted by the French army. Others had ended up at the ETVS after a more mysterious route: weapons brought back by military attachés or by intelligence officers posted abroad, or weapons seized from the depots of subversive organizations or from ships delivering contraband. Although the archives remain discreet concerning the provenance of these weapons, some expressions in the official reports have a whiff of adventure and mystery about them: "Weapon brought back by an officer of the second office," "Drum magazine from Suomi brought back from Finland by an officer," "Submachine gun seized on board a Japanese vessel at Bordeaux," etc.

The war in Spain also brought some new submachine guns to the ETVS, which subsequently joined the weapons captured from the enemy in combats that took place during the Phoney War in 1940.

Let us remember the main tests carried out by the ETVS on foreign submachine guns (at least those that we have managed to trace).

WEAPONS OF THE FIRST WORLD WAR

The German MP 18/Is captured in the last

Thompson model 1923 SMG; this version was equipped with a bipod and a stock less sloping than on the model 1921, and it was designed more to be used as a support weapon in the manner of a light machine gun. Subject to order, it could be delivered in 45 ACP, 7.63 mm Mauser, 45 Remington Thompson, or .451 SL caliber. This version did not experience much success, and only a few were made. *Extract from an Auto-Ordnance company manual*

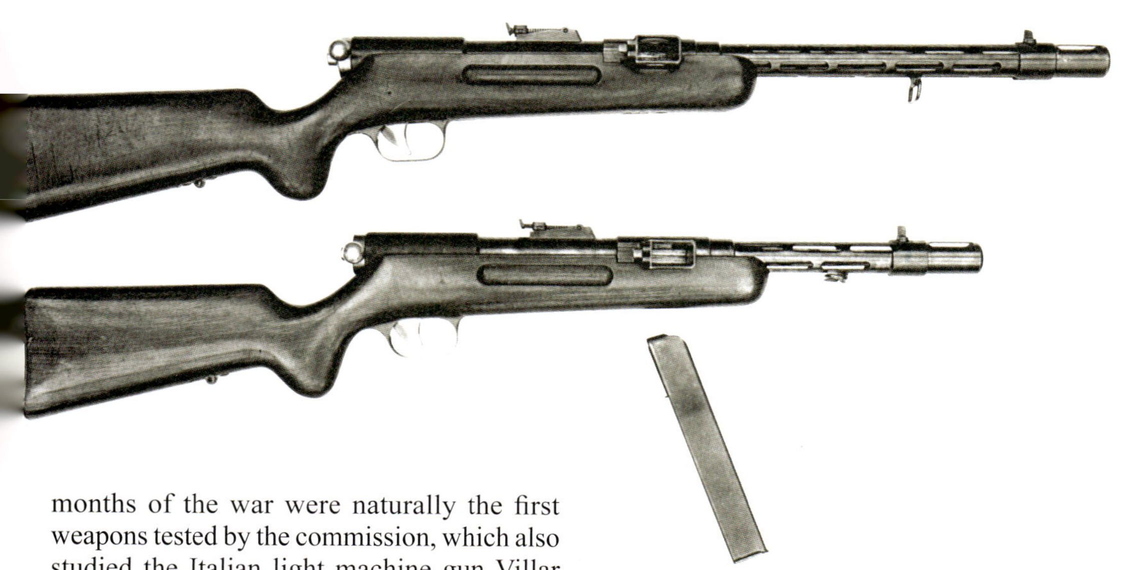

A long Bergmann submachine gun no. 4427 and a short Bergmann submachine gun no. 3466 were given to the ETVS on August 18, 1938, for evaluation by the 2nd bureau SR (the French intelligence service of the period). *CAA*

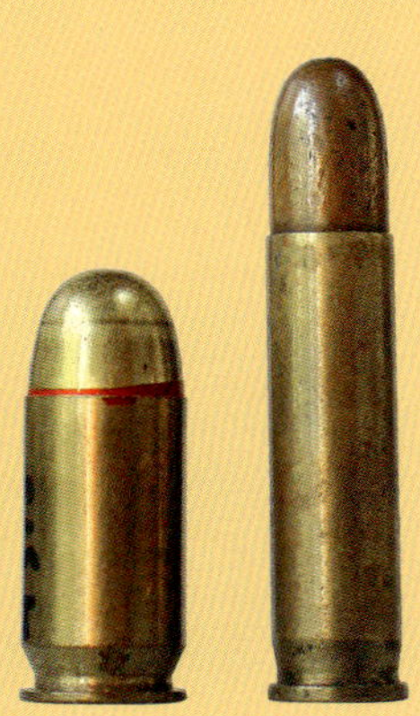

LEFT: a standard .45 ACP caliber cartridge (here a Second World War make with steel case), used in the various models of the Thompson submachine gun. **RIGHT:** the .351 SL cartridge for which the Thompson prototypes were tested. *Loïc L'Helguen*

months of the war were naturally the first weapons tested by the commission, which also studied the Italian light machine gun Villar Perosa and the first Italian submachine guns derived from it. The specifications of the submachine gun contained in the 1921 program were very clearly influenced by the characteristics of the MP 18/I (generally designated in French documents of this period under the name of "Bergmann submachine gun").

The French STA submachine gun is based directly on the German weapon, although its magazine was borrowed from the Italian army (designated in the test reports under the name "Revelli submachine gun").

Schmeisser MP 28/II Submachine Gun

In 1933, the ETVS tested the improved version of the MP 18/I: the Schmeisser MP/II. This was a new weapon, supplied with three interchangeable barrels in 7.63 mm Mauser and 7.65 mm and 9 mm Parabellum caliber. Information that has been verified states the purchase of a small quantity of Schmeisser MP 28/II-type submachine guns by France during the 1930s. These were weapons assembled by the Pieper factories of Liege from parts made in Germany by Schmeisser.

It is possible that this order, if indeed it really did take place, was destined for the police and not the army, which explains why there is no trace of it in military archives.

Thompson Submachine Gun

Between 1921 and 1927, several versions of the Thompson submachine gun were presented by the American Auto-Ordnance company, created by Gen. J. T. Thompson, to market his weapon. From 1921, the French army tested the new Thompson submachine gun in 11.43 mm caliber (.45-caliber ACP). The Versailles commission for experiments had wisely considered that the use of pistol ammunition did not justify the presence of a delayed-opening mechanism. In addition, the .45 ACP ammunition was of no particular interest when compared to the 9 mm Parabellum that the French army had decided to use in the submachine guns tested while waiting for a definitive choice of a cartridge common to future submachine guns and automatic pistols. New trials were carried out in 1924, with a sample in .45 ACP and another sample fitted with a "Browning type" bipod and chambered for .45 Thompson ammunition. Although the accuracy of the weapon in single-shot firing of .45 Thompson was superior to that of the sample in .45 ACP, the vibration caused by full-auto fire on the .45 Thompson rendered firing extremely inaccurate. The commission observed that only the use of a carbine ammunition could justify the delayed-opening mechanism on the weapon, and that it wished to be presented with a sample in .351 SL, also proposed by Thompson.In 1926,

Two German submachine guns: an MP 18/I with modified magazine housing for the use of straight magazines, placed on an MP 28/II. Note the presence at the rear of the magazine housing of the MP 18/I, of a "universal" safety, often added by the police on its weapons. *Marc de Fromont*

Italian 38a model Beretta submachine gun

The semiautomatic Beretta 18/30 model carbine tested in January 1939 by the Infantry Testing Commission (CEI)

a comparative test was carried out between a .45 ACP Thompson and another in .351 SL caliber, delivered with 1,000 cartridges. Various firing incidents (broken extractor, broken locking parts, damaged cleaning flannelette) caused the test to be stopped, although the modest quantity of ammunition accompanying the weapon did not allow the trials to be very thorough.

In January 1927, the research and test inspector of the artillery instructed the ETVS to test a Thompson submachine gun in 9 mm caliber that had been brought back from the United States by the French military attaché. This weapon came with a proposition for the assignment of rights to manufacture the weapon under license in France. Although it functioned using the same principles as the 1921 models, this Thompson submachine gun in 9 mm caliber had a different appearance, as found on some prototypes that the British firm BSA planned to produce around the same period. The firing system, the trigger, the angle of the unlocking rail, and the rear sight of this model were slightly different from those on the 1921 models and their derivatives that had been tested up to that point by the ETVS. The very high rate of fire (1,200 shots per minute) severely harmed the accuracy, to such an extent that it was stated that groupings in short bursts at 50 meters were the equivalent to those made by the French STA submachine gun at 100 meters! Taking up the conclusions of previous tests carried out with the Thompson submachine gun, the testing commission decided that there was no need to proceed with Thompson's proposals.

Bergmann MP 35 Submachine Guns

On February 13 and 17, 1940, the ETVS also tested twelve Bergmann MP 35s in 9 mm Parabellum caliber seized at Bordeaux from the Japanese ship *Kasima-Maru*. The commission confirmed the satisfactory operation of two weapons tested from the batch and concluded that "the seized weapons are likely to be used": the question is, were they?

It should be noted that two years earlier, tests had already been carried out on a long Bergmann submachine gun no. 4427, a short Bergmann submachine gun no. 3466, and a Beretta automatic carbine.

Carbines and Beretta Submachine Gun

In 1938, the Beretta semiautomatic carbine model 18/30, imported into France by the Seytre establishment of Saint Etienne, had been tested at Mourmelon by the CEI in 1938. In May 1940 (just before Italy declared war on France), a model 1938 Beretta submachine gun, no. 3162A, submitted to the ETVS by a representative of the Beretta company, was also tested in order to determine if it could be used by the armies.

If not for the risk of unexpected firing, the Beretta submachine gun was considered "likely to become a good weapon of war. Its accuracy and resistance are satisfactory, its firing, handling and maintenance render it easy to use."

The Spanish Labora submachine gun

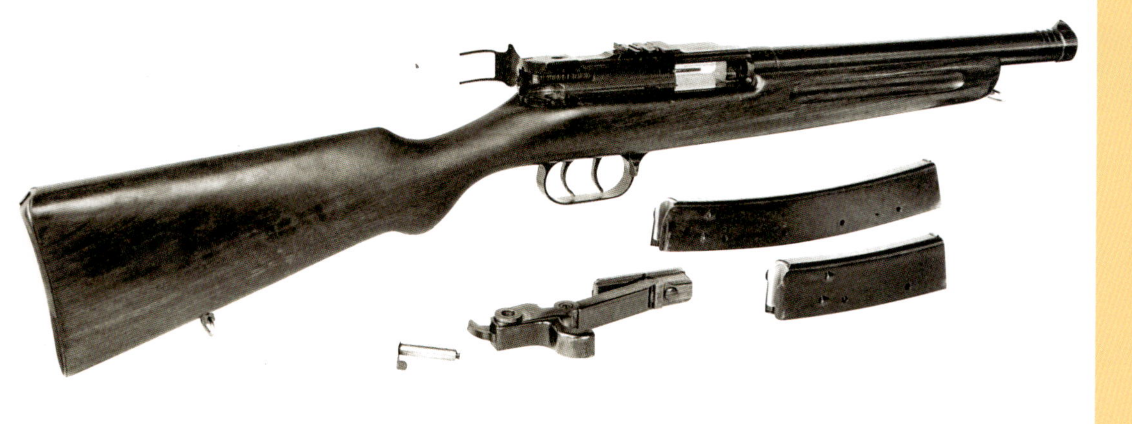

The FN Herstal submachine gun with a toggle lock system researched by ETVS

Weapons from Spain

Prior to the outbreak of the Spanish Civil War, the Spanish company Star presented two of its submachine guns to the ETVS, one in .38 automatic and the other in 7.65 mm caliber Long. The fact that this company had taken the trouble to produce a weapon in this caliber used solely by France is testimony to the hope of seeing its submachine gun adopted in that country.

The Star submachine gun trials had no follow-up, since Spain, already broken by violent social disruption, headed into a cruel civil war lasting from July 1936 to March 1939.

Various types of locally and foreign-made submachine guns were used by both camps during the Spanish Civil War. The Republican forces, whose weaponry had been supplied by international traffickers, were using weapons of an extremely disparate nature, supplied throughout the civil war by intense international trafficking. Among these, almost all models of submachine guns available on the market up to 1938: Erma-Vollmer, Bergmann MP 35, Schmeisser MP 28/II and its Spanish copy the PL FAI, Labora, Star, Finnish Suomi, Estonian Tallin, and the Swiss SIG 1920 (copy of the MP 18/I).

Many of these weapons were deposited at the border in the Pyrenees when factions of the Republican army that had been able to escape from the destruction and capture sought refuge in France after the victory of the nationalists. The submachine guns retrieved on that occasion were stored at Clermont-Ferrand, and some specimens of the different models retrieved were tested by the technical units of the French army on the eve of the war. As will be seen in the following chapter, a new evaluation of those models retrieved in sufficient quantities was carried out from the end of 1939 to the beginning of 1940, in order to determine which would be put back into service in French combat units.

Evaluation of Captured German Weapons

Between the outbreak of hostilities in September 1939 and the collapse in May 1940, various submachine guns were captured by French combatants. The majority of them were directly reused by French troops. A report on the test of a Steyr Solothurn submachine gun with the number 4486 was, however, found in the archives of the arsenal. This weapon, qualified by the ETVS as a "model unknown until now," was deformed by the impact of a bullet on the muzzle and had a broken frame and could not therefore be tested. The report states the weapon was chambered in 9 mm Bergmann (in reality, it was more likely 9 mm Steyr). Two months later, a new specimen of the Steyr submachine gun in 9 mm Mauser with the serial number 4812 arrived at the ETVS, and this time the weapon was in firing condition and could be tested.

It should be stressed that at the beginning of 1940, the production of the MP 38 was still very low, and this type of weapon was not commonly used by German troops; only the parachutists had received them as a priority. Photographs of the period show the German troops armed principally with modified 18/I, MP 28/II, Erma Vollmer, Bergmann 35, and Steyr Solothurn MP 34 types, which had become part of the arsenal of the Wehrmacht after the annexation of Austria.

The Steyr Solothurn MP 34(ö) submachine gun, a weapon of Austrian origin that became part of the arsenal of the Wehrmacht after the Reich annexed Austria in 1937. This weapon, captured during combat and with a deformed muzzle and split stock, was examined by the ETVS in 1940. Firing tests were carried out when a second, intact specimen was captured later.

CHAPTER 6

THE SECOND WORLD WAR

French infantrymen in the South of France at the end of June 1940, showing their Erma-Vollmer submachine guns. *GU*

At the beginning of the war, as the German and French armies occupied sedentary positions opposite one another, fighting had unexpectedly become a series of skirmishes between reconnaissance patrols in the unoccupied zone separating the positions.

In this type of combat, the MP 18, MP 28, Erma Vollmer (EMP), and Bergmann types used by the Germans proved to be particularly effective since they substantially increased the firepower of the patrols, without the bulkiness and the weight of the light machine gun.

The Petter submachine gun had been adopted as the regulation weapon by the French forces under the name "Pistolet Mitrailleur 7.65 mm L modèle 1939 Petter," or Petter model 39 submachine gun, but since the production capacity of the SACM factories was entirely taken up with the manufacture of the PA 1935 model on the declaration of war in 1939, it was not conceivable that the mass production of the Petter submachine gun could begin within a reasonable period. A letter dated September 9, 1940, from the director of the MAS and addressed to the mechanical manufacturing department of the Armaments Ministry, stated that if the mass production of the Petter submachine gun had been ordered on January 1, 1940, the first weapon would not have been delivered until July 1941. MAS, on the other hand, would have already delivered 35,000 MAS submachine guns by that date.

April 1, 1940, in Lorraine, the Corps Francs of the 73rd Infantry Regiment partially equipped with the Erma Vollmer submachine gun

The general staff requested that the weapons factory at Saint Etienne (MAS) start manufacture of a modified version of the SE MAS model 1935 as quickly as possible, on which the safety placed on the left side of the receiver was removed in favor of a new safety operated by the forward folding of the trigger.

In parallel, the technical services of the army were eager to research all the possibilities for the supply of weapons of foreign origin.

FOREIGN SUBMACHINE GUNS OF THE FRENCH ARMY IN 1940

These weapons came from several sources: Spain, Finland, Belgium, and the United States.

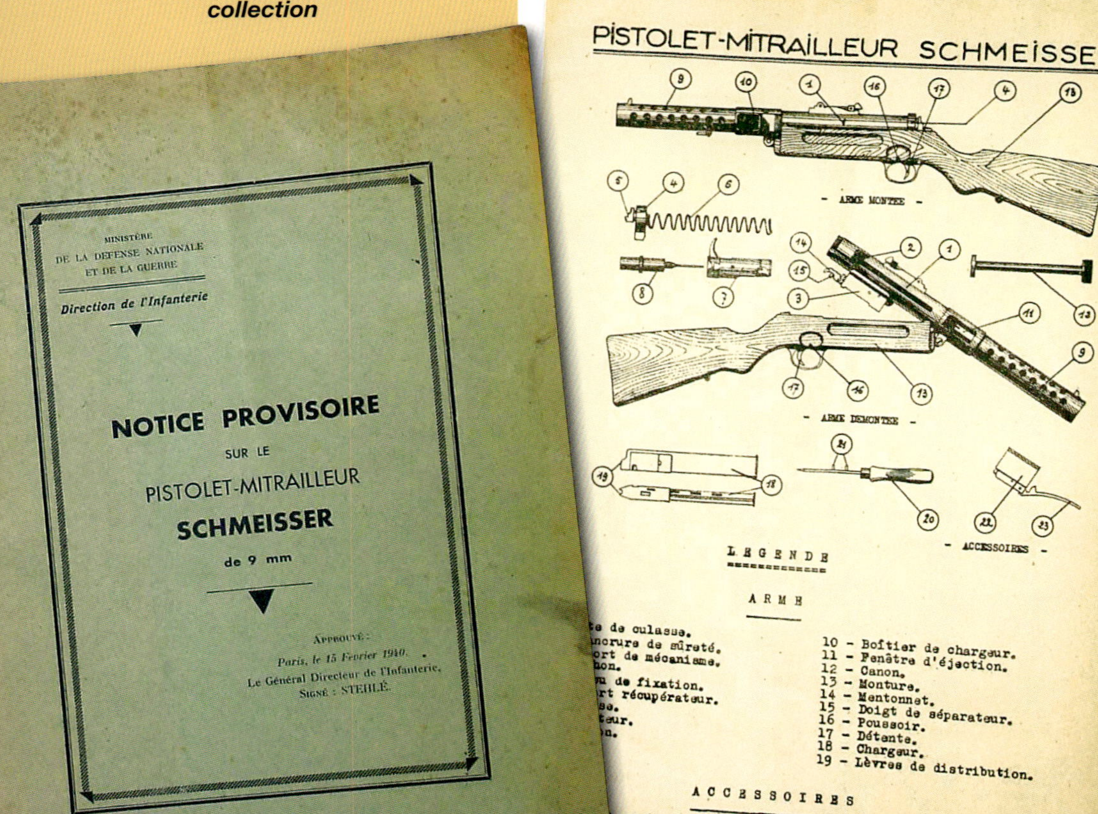

MINISTÈRE
DE LA DEFENSE NATIONALE
ET DE LA GUERRE

Direction de l'Infanterie

NOTICE PROVISOIRE
SUR LE
PISTOLET-MITRAILLEUR
SCHMEISSER
de 9 mm

APPROUVÉ :
Paris, le 15 Février 1940.
Le Général Directeur de l'Infanterie,
SIGNÉ : STEHLÉ.

PISTOLET-MITRAILLEUR SCHMEISSER.

- ARME MONTÉE -

- ARME DÉMONTÉE -

- ACCESSOIRES -

LÉGENDE

ARME

10 - Boîtier de chargeur.
11 - Fenêtre d'éjection.
12 - Canon.
13 - Monture.
14 - Mentonnet.
15 - Doigt de séparateur.
16 - Poussoir.
17 - Détente.
18 - Chargeur.
19 - Lèvres de distribution.

ACCESSOIRES

French notice relating to the Schmeisser 28/II MP. The cover bears the name of 2nd Lt. Calame, group leader of the 21st BCP Corps Francs. *Private collection*

Labora 1938 submachine gun no. 665

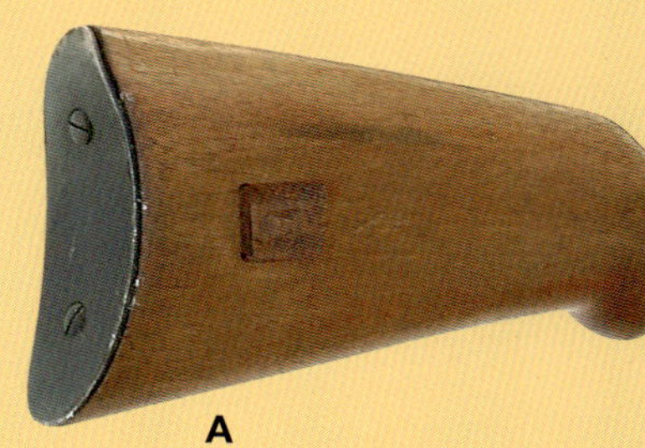

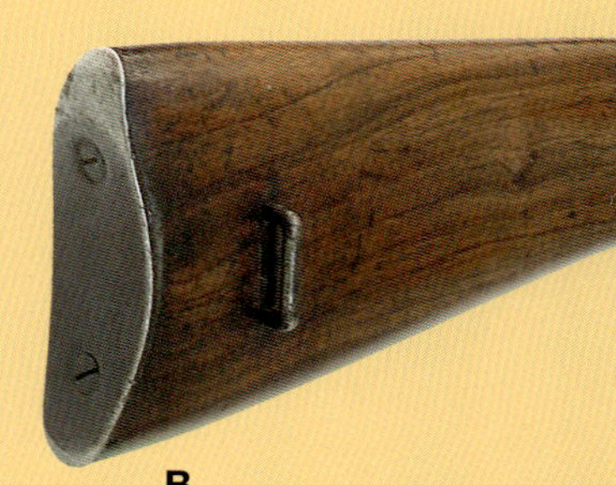

Spanish Submachine Guns

When the situation rendered the supply of submachine guns within frontline units imperative, ETVS researched the possibility of using Spanish submachine guns stored in the regional equipment maintenance and repair center at Clermont-Ferrand.

The Labora and FAI (Spanish copy of the Schmeisser MP 28/II) submachine guns were available only in restricted numbers. In addition, both these weapons used regulation Spanish 9 mm Bergmann Bayard (9 × 23 mm) ammunition. They both were eliminated due to the difficulty of procuring a supply of ammunition.

The armament center at Clermont-Ferrand had in its possession, however, 3,250 Erma-Vollmer (EMP) submachine guns, chambered in 9 mm Parabellum caliber. Significant stocks of this ammunition, captured in 1918 from the German army, are still held at the Miramas depot.

Although there were only a limited number of magazines, these EMPs could provide valuable additional weaponry to the Corps Francs. Previously, their operation had to be verified by French technical services. At the end of 1939, three EMP submachine guns formerly retrieved from the Spanish border with France were sent to the ETVS for evaluation by the center at Clermont Ferrand.

The ETVS also tested a fourth EMP that the report designated as a "prototype made by the Swiss firm EMP."

The commission concluded that "the army's use of the Erma-Vollmer pistols stocked

Comparison between a German EMP (*A*) and a Spanish-type EMP (*B*) used by the French Corps Francs. Apart from the differences between the shape of the actuator knob and the fixing of the sling to the stock, the German weapon is marked "EMP," and the fire mode selector positions are identifiable by the letters "E" and "D." The Spanish model is in principle without marking, and the selector is identifiable by the letters "T" and "A." In these photos, the German EMP is equipped with a notched, flip-over-leaf rear sight, but some examples with a tangential sight were also used by the German army. This specimen has a rotating safety added to the receiver at the request of the German police force. This part was sometimes absent on the EMPs used by the Wehrmacht or the Waffen-SS.

France, May 1940, a member of the 3rd SS Panzer Division armed with an Erma Vollmer submachine gun

in the regional center at Clermont Ferrand can be envisaged, subject to careful prior verification of the weapons and magazines that will be sent."

However, the case of the EMP presented as of "Swiss origin" remains a mystery, and the conclusion of the test commission was quite harsh in regard to this: "As to the proposal of the Swiss firm to supply weapons of this model, it must be considered for the moment with a certain reserve. This firm has never made pistols in 9 mm caliber, and the prototype it presented was not in a condition to function properly. In addition, the thermal treatments applied to certain parts appear to be incompatible with their good resistance, which would suggest that the firm is unfamiliar with the manufacture of weapons."

A new test carried out several months later, after recovery and repair of the weapon by the manufacturer, was no more successful. The test report was dated May 31, 1940, and was much too late for the EMP.

The EMPs from Spain were put into service at the front, and several photos of the period show them in the hands of men of the Corps Francs. Since only 1,540 magazines were available, it can be concluded that only 500 to 700 EMPs were finally sent to the front, a minimum of two magazines per weapon seemingly the minimum necessary.

In his book *Corps Franc*, Albert Merglen gives many interesting details concerning the weapons used by its men. He points out in particular the fact that the first EMPs were issued in his Corps Franc group around March 1940: "An old dream of the Corps Franc comes true: there are some German model submachine guns that must come to Spain, and boxes of 9 mm cartridges, with German inscriptions and the date of manufacture: 1917. Excellent weapons, easy to handle, light, and with rapid firepower in the action of surprise, they certainly cannot replace a carbine at four hundred meters but are indispensable in this hand-to-hand fighting in the woods."

Later on, the author tells us his Corps Franc group of seventy-eight men is equipped with nine submachine guns and seven light machine guns; the rest of the armaments were essentially composed of carbines, along with knives, grenades, and pistols. During engagement with an enemy patrol, the Corps Francs chasseurs noted that one man in three of the *Stosstruppe* was armed with a submachine gun.

In his book *Sacrée drôle de guerre*, Saint-Roc also refers to a Spanish submachine gun seen by his Corps Francs group: "The group has been issued with an SMG taken near Argelès from the Spanish Republican weapons

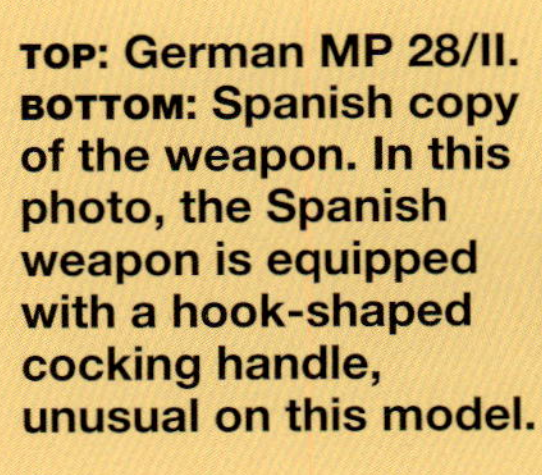

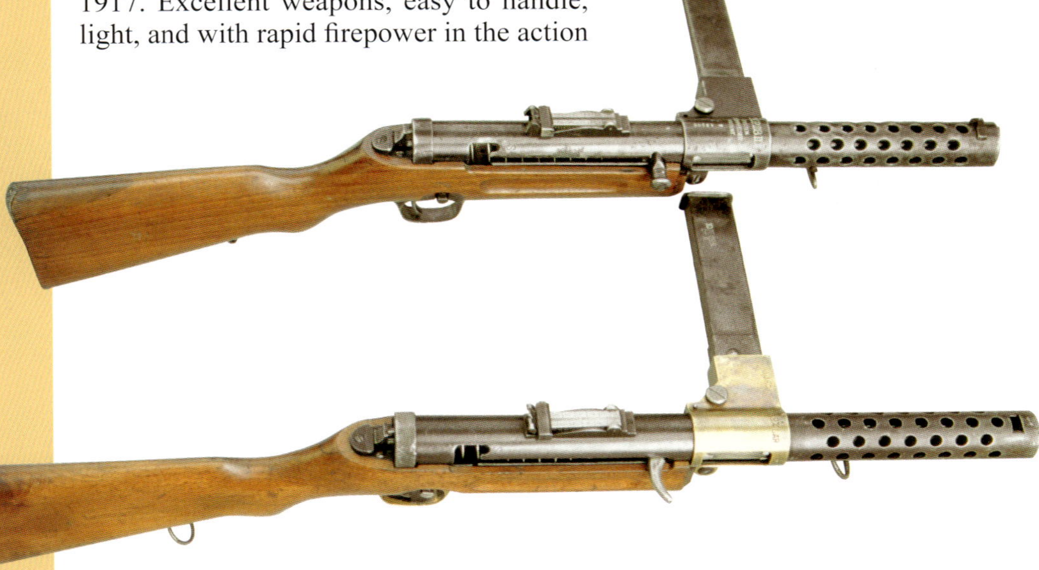

TOP: German MP 28/II. **BOTTOM:** Spanish copy of the weapon. In this photo, the Spanish weapon is equipped with a hook-shaped cocking handle, unusual on this model.

Two Finnish Suomi submachine guns: a standard model KP 31 with a protective barrel jacket with angled extremity, and a KP 31 SJR, recognizable by its barrel extended by a lift-up compensator. Note the typical appearance of the stock in birch wood on this weapon.

stocks. It was very worn, had no markings, and fired somewhat oxidized German cartridges. But with a little care and a lot of petroleum, everything worked to perfection, and the extra automatic weapon gave a boost of confidence. This SMG must be German, since the cartridges are Nazi. Rather than going through the Spanish Republicans, we might just as well remove the middleman and buy directly from Jerry. They would give us a good price and sell us new weapons!"

Some EMP weapons captured from French troops in 1940 were put back into service by the Germans for their own purposes, under the name MP 720 (f). Several specimens rescued from the mayhem would reappear in the hands of the Milice. The only known photo was taken at the chateau of Uriage, at the Milice officers' school. All photos of the Milice in operation show them armed with Sten submachine guns recovered from air dropping, which had been captured from the Resistance. It is therefore likely that the few EMPs used at Uriage were weapons destined for the instruction of officers of the Milice, pending the allocation of a sufficient number of Sten.

Finnish Submachine Guns

It seems that a hundred or so Finnish Suomi submachine guns in 9 mm Parabellum also retrieved from the Spanish border were also sent to the armies. In his reference book *Weapons of the French Infantry, 1918–1940*, Stéphane Ferrard mentions the existence of 300 Finnish Suomi M/31 submachine guns in the depot at Clermont Ferrand, 150 of which would be selected by Col. F. Toulouse (ER). This officer went on to carry out test firings at −25°C before sending this batch to the front.

On February 17, 1940, a Suomi submachine gun no. 6402, with a drum magazine that had come from Finland, was also tested, and the report indicates that it had been given to the ETVS by an officer of the EMA (Etat-Major des Armées, or army general staff). The ETVS concluded that "the drum magazine is roughly the same weight as three parallelepiped magazines. In one case there are seventy-two cartridges for 1.935 kg, and in the other, ninety-six cartridges for 1.960 kg. The drum magazine therefore has no real advantage."

Belgian Submachine Guns

In his research dedicated to French submachine guns of this period in the Gazettes des Armes under the title "Mitraillettes et Bourguignottes," François Vauvillier mentions the statement made to the commission of the Assemblée Nationale tasked with investigating events in France between 1933 and 1945. The statement was made by the engineer General Happich, the director of weapons manufacture from January 1936 to April 1939, then weapons manufacture inspector until August 20, 1940, reporting the purchase of 1,000 Thompson submachine guns and 1,000 Schmeisser submachine guns.

Young member of the Milice during training. This man is armed with a Schmeisser MP 28/II submachine gun. The Germans, having always been reluctant to arm the Milice, first equipped themselves with recovered French weapons. The first submachine guns of the Milice were most likely EMPs formerly used in the Corps Francs. The use of this MP 28/II is more exceptional; it is highly probable that this Schmeisser submachine gun came from a batch bought by France in Belgium in 1940. Later, the Milice would be equipped mostly with British weapons taken from the parachute drops captured from the Resistance. *DR*

The Schmeisser submachine guns were made in Germany starting in 1928 by the Haenel firm of Suhl, under the name "MP 28/II." As has been discussed previously, an example of this weapon had undergone comparative testing with an MP 18/I by the ETVS in 1933. Although Haenel was authorized to sell its production to German police forces, exportation remained forbidden by the disarmament clauses in the Treaty of Versailles. Haenel made a deal with a well-known Belgian arms manufacturer—the former Pieper establishments in Liege—for them to ensure the assembly of parts made in Germany and also to market the MP 28 on the global market. The French army placed an order for 1,000 Schmeisser submachine guns from this Belgian company.

Colonel Bacque, a French officer, was sent to Belgium to accept delivery of the first 300 submachine guns, accompanied by 900 magazines, 300 loading devices, 300 universal tools, 300 cleaning rods with two chamber brushes, and 600,000 9 mm Parabellum cartridges. It is not known if the rapid deterioration of the situation in Belgium allowed for the delivery of the remainder of this order. In fact, a letter dated April 8 mentions the absence of equipment for these weapons.

Berthier rifle sling rings added by the French army on its Thompson submachine guns

American Submachine Guns

The Thompson submachine gun was one of the only weapons of this type available on the market in 1939, when the French army was faced with an urgent need to equip its frontline troops with submachine guns, until such time as when manufacture on a national scale was set up. Despite the reservations made during the previous years concerning the adoption of this weapon, France placed an urgent order with the United States for 3,000 Thompson 1921s.

A second order for 3,000 Thompsons was also planned but was never delivered. It is known that when Russel Maguire took control of the Auto-Ordnance firm, there remained 4,700 unsold Thompsons out of the initial production of 15,000 weapons made by Colt in 1920 and 1921. If the 3,000 samples ordered by France in 1940 are subtracted from this figure, not enough remained to satisfy a second order for 3,000 weapons.

Russel Maguire, having predicted a huge increase in demand for submachine guns, entered into a contract with the Savage factory in December 1939 for Savage to restart manufacture of the Thompson submachine gun in its 1928 version. In his book *Thompson: The American Legend*, Tracie L. Hill states that the first 203 Thompsons were completed by Savage in April 1940. The second order for 3,000 Thompsons that the French army had planned to make with the Auto-Ordnance was therefore never delivered. It is likely that these weapons were subsequently bought by the British government.

The ETVS no. 3 secret report, dated February 1, 1940, mentions precision tests carried out on two Thompson submachine guns taken from a batch of weapons delivered to France along with a batch of cartridges:

- weapon no. 10384, with no compensator
- weapon no. 10171, with a Cutts compensator

Thompson model 1921, serial number 12551, from the order made by France to the United States in 1940

The report specifies that the weapons were accompanied with an actuator and a spring, meaning the rate of fire could be reduced (clearly the model 1928 spring and actuator, which Auto-Ordnance offered to mount as an option on its 1921 models, along with the Cutts compensators). The commission concluded that despite its weight, the Thompson submachine gun was easy to handle and to shoulder and was easily disassembled without tools. Full-auto fire could sometimes be difficult for shooters of small build. The significant muzzle rise in full-auto fire of the model without compensator was highlighted.

When the model 1928 spring and actuator were mounted on the weapon, the rate of fire was reduced from 920 to 750 shots per minute, and the weapon became easier to use and more accurate during full-auto fire.

In conclusion, the weapon operated satisfactorily, but its accuracy was average. If a new batch of weapons had to be ordered from the United States, it was recommended that models fitted with compensators, actuator, and springs ensuring a slower rate of fire be chosen. The commission criticized the fact that the weapon was not fitted with sling rings, and several months later it was requested that rotating sling rings of the type mounted on Berthier rifles and carbines must be fitted on the left side of the pistol grip forward and under the stock.

SECRÉTARIAT D'ÉTAT A LA GUERRE

ÉTAT-MAJOR DE L'ARMÉE

DIRECTION DE L'INFANTERIE

NOTICE

SUR LE

PISTOLET-MITRAILLEUR THOMPSON

de 11,25 mm., Modèle 1921

Approuvée par le Ministre Secrétaire d'Etat à la Guerre, le 24 mars 1941

CHAPTER 7

WAR PRODUCTION MAS-35 SUBMACHINE GUNS

SE MAS-35 submachine gun. We have chosen to name this "wartime" to differentiate it from the real prototypes of 1935. These weapons came from the small batch made by the test section of the MAS in 1940 to supply submachine guns to combat units, until the mass production of the MAS model 38 submachine gun was begun. Weapon number F 504 is seen here. The safety lever placed on the left of the receiver, which existed on the test models, has been removed and replaced by a safety device operated by the forward movement of the trigger.

Starting in 1924, the multiple prototypes of the MAS submachine gun presented to the ETVS and various commissions of the French army were made by the test section of the MAS. That is why approximately 300 STA submachine guns were made by this division around 1925, and around fifty MAS model 1935 submachine guns later.

The equipment of the "test section" did not allow for the transition to large-scale production, which was the responsibility of the "production section" of the MAS. The production section could be set up only after planning by the "method section."

Mass production involved the following:

- issuing plans
- allocating personnel, premises, machine tools, and raw materials to produce the new type of weapon
- adjusting instruments of measurement and verification, developing manufacturing, and revenue control procedures
- manufacturing batches of spare parts
- setting up a maintenance-and-repair policy
- writing and distributing instructions for use of the weapon
- producing or subcontracting out the associated equipment (magazine pouches, cases, accessory pouches, etc.)

Even using emergency procedures, these diverse operations required several months before the first mass-produced weapons could come off the assembly lines. This timeline was further lengthened by the war situation, which mobilized the resources of the MAS to produce its model 1936 rifle.

To deal with the matter urgently, the test section, whose research activities were temporarily suspended, was ordered to produce 500 SE MAS-1935 submachine guns, using its own resources. These weapons were modified by the removal of the safety lever placed on the left side of the frame, and by the addition of a safety controlled by the forward folding of the trigger.

OPPOSITE PAGE: MAS-35 made in 1940, with objects evoking its use during the *Sitzkrieg* ("Phoney War").

Man of the GRCA (Groupe de Reconnaissance de Corps d'Armée), armed with an SE model 35 submachine gun "somewhere in France," probably in June 1940. Two other modern items of equipment in service in the French army can also be seen in this photo: an R40 tank and a 25 mm antitank gun. *Philippe Thiry*

CAL
65L
Type SE-MAS 1935 F.935

A. **The sight was made up of two leaves bearing an eyesight for distances of 100 and 200 meters, respectively. These leaves could be folded during transport, and they were able to sink entirely into the receiver, leaving a smooth surface.**
B. **Rear leaf (100 meters unfolded)**
C. **200-meter leaf unfolded**

The receiver of these weapons was marked "PM SE-MAS-1935 cal. 7,65 L." The initials "SE" indicate that the weapons were produced by the "section d'essais" or test section, and there are discs of the MAS on the stock dated from March to May 1940. According to the excellent study by Daniel Mastier and Stéphane Ferrand on the MP 38 submachine gun in issue no. 66 of the *Gazette des Armes* (December 1978), and to the study by François Vauvillier mentioned above, the SE MAS-35 submachine guns made in 1940 equipped tank units, including the 347th CACC, destined for Norway, along with batteries of 47 mm self-propelled guns at the rate of one submachine gun per vehicle.

From the outbreak of war, German troops had been equipped with submachine guns. At that period the production of the MP 38 had just started; therefore, perhaps the models used by our adversaries were instead the Bergmann MP 18/I, Schmeisser MP 28/II, Erma Vollmer, Bergmann MP 35/I, and Steyr Solthurn MP 34. These were weapons made in limited numbers between the wars for police forces and for export, and for which the German army had requisitioned the available stocks at the beginning of the war.

On the French side there was a period of great deprivation up to the end of 1939, during which the French patrols complained of frequently finding submachine guns in the hands of enemy assault troops when they themselves were lacking any decent firepower.

Comparisons under different angles of the grips of 35 (*A*) and 38 (*B*) submachine guns

From the last months of 1939, the progressive supply of the aforementioned foreign submachine guns to French forces, along with the arrival of the first SE MAS-1935s, meant the Corps Francs were adequately equipped with weapons. In May 1940, the number of weapons allocated was probably similar on the German and French sides, and the disparity of the models in service was equivalent on both sides. All the more so since the "test section" would finally be able to manufacture not 500 but 1,958 SE MAS-35 submachine guns, before the mass production line for the MAS-38 started operation in May 1940.

Model 38 submachine gun, serial number F-2005, bearing the date June 1940 on the stock. Even though marked PM 38, this first example has kept a receiver similar to that on the MAS SE 35 submachine gun. ***Musée du souvenir, collection Klamerek***

MAS-38 SUBMACHINE GUN

During this time, the "production" section of the MAS urgently set up the mass production of the new MAS-38 submachine gun. This was to make the 19,500 model 38 submachine guns, the order it had received from the War Ministry in January 1939, as quickly as possible.

This new weapon, differing only in a few details from the SE MAS-35, was adopted on May 9, 1940, under the name "MAS model 38 7.65 mm caliber Long." The obscure reasons that governed the choice of this year for this submachine gun are not known, even though it was adopted after the Petter, which had been given the name "Model 1939 submachine gun."

The first samples of the 38 began leaving the MAS on a small scale in May 1940, only a few weeks before the armistice of June 22, which stopped production for a time.

Several months later, the MAS, which had been placed in an unoccupied zone by the armistice agreements, started production of the model 38 for the army of the armistice. It is estimated that 14,000 weapons of this type were made before the German invasion of the unoccupied zone in November 1942.

Although a few of these weapons appear on propaganda photos of the 10,000-strong armistice army that had been authorized by the Germans to be conserved in a free zone, the majority of them most likely went to North Africa and were used to equip French units that had taken up arms with the Allies after 1942.

After the occupation of the "free" zone in November 1942, the German army took control of the MAS and had production of the MAS-38 submachine gun continued for its own use, and it was integrated into its arsenal under the name MP 722(f).

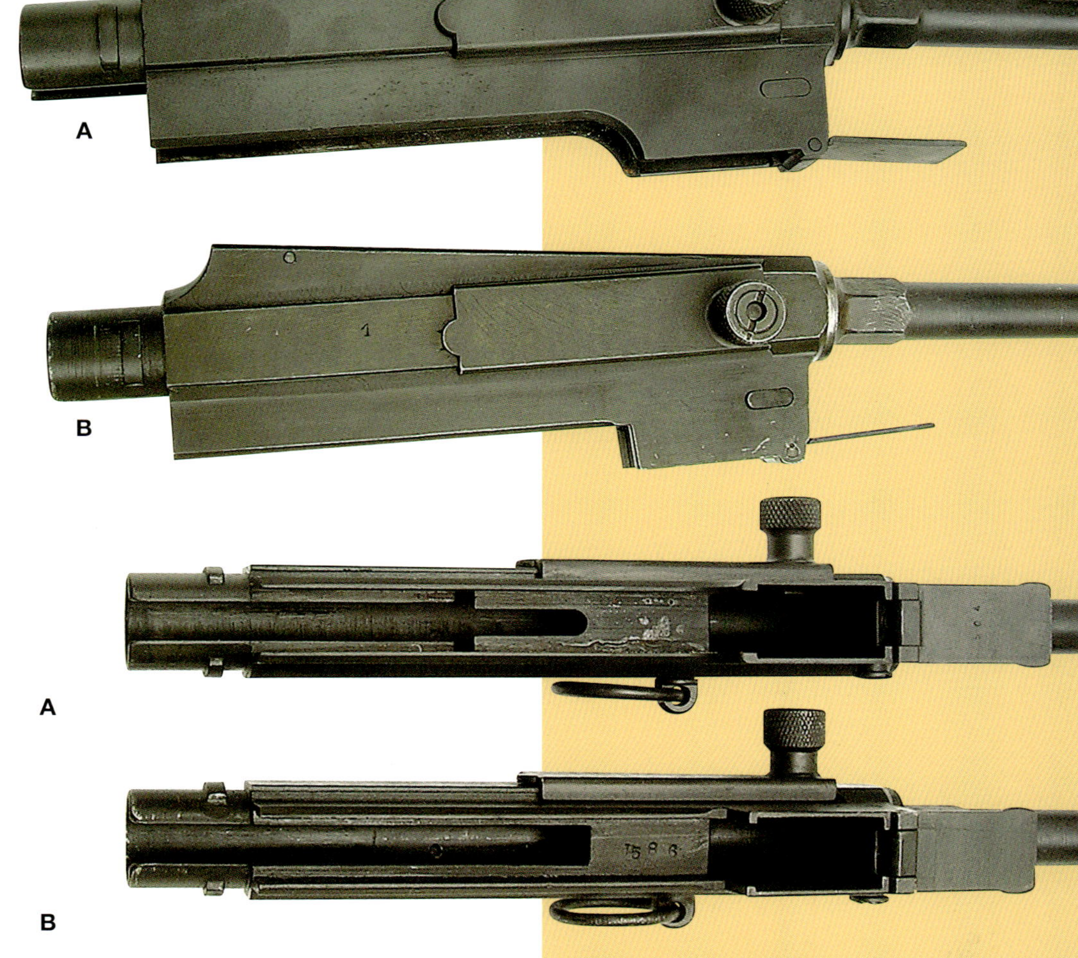

Comparison between the receiver of a 35 (*A*), which was conserved on the first 38 (*B*) submachine guns made in 1940, and one on a 38 as made from 1942

The production of the model 1938, deliberately slowed down by the workers of the MAS under German occupation, became large scale starting in November 1944, very shortly after the liberation of Saint-Etienne. Only 30,000 weapons were made from 1940 to 1944, while 39,000 were made during a single year in 1945.

The subsequent career of model 38 within the French army will be seen later.

CHAPTER 8

THE SUBMACHINE GUN IN FRANCE FROM 1944 TO THE PRESENT

Indochina, Christmas night 1951, an observation post with one of the soldiers using an MAS-38

FRENCH WEAPONS

A document dated June 1946 from the DEFA (Weapons Research and Production Unit) summarized the criticisms coming from users concerning the MAS model 38 submachine gun.

These were contained in three main points:

- the absence of a folding stock
- the absence of a firing safety
- The use of specific ammunition not interchangeable with that of numerous other weapons in 9 mm Parabellum caliber or .45 ACP used at that time by the French army. Contrary to what many collectors believe today, it is not the lack of firepower of the 7.65 mm Long ammunition that was questioned, but rather its lack of interchangeability.

From the Second World War onward, the characteristics of a new submachine gun destined to replace the MAS model 38 and the foreign submachine guns used by the French army were stated in specification programs of June 5, 1945 (note 27 798 ST/ARM), October 28, 1946, and May 19, 1948 (DM 31 474 ST/ARM).

The French army had also opted to adopt the 9 mm Parabellum as the sole ammunition for its future submachine gun and its future automatic pistol (decision of July 20, 1946).

After 1945, the weapons services underwent restructuring. The CEI was dissolved and two organizations for verification of the weapons remained:

- the ETVS for small-caliber weapons
- the technical establishment at Bourges (ETBS) for the artillery (caliber greater than 20 mm)

During the 1970s, the ETVS was dismantled and all experimentation and research were transferred to Bourges within the ETBS, which separated its activities into small-, medium-, and large-caliber sections.

Another organization played a role in the definition of weapons: the technical section of the land army (STAT), successor to the technical section of the army—the STA, initially set up in the 6th District of Paris at Place Saint Thomas d'Aquin, with a branch at Satory.

MAS-39 submachine gun made after the Liberation, seen with accessories evoking the immediate postwar period and the time when the first expeditionary corps units were sent to the Far East. ***Marc de Fromont***

Stock disc

MAS model 1938 made in 1948. Phosphate finish and beech stock.

Serial number preceded by the letter "g"

MAS-38 transformed in 9 mm caliber Parabellum, developed by the MAS after the war. This weapon uses the STA 1924 magazine, with a stopping spacer slightly reduced in thickness. *Jacques Barlerin*

A sailor at the start of the war in Algeria observing the port of Alger, armed with his MAS-38 submachine gun. *GU*

Prototype of the MAS-38 in 9 mm caliber Parabellum, with a thread at the extremity of the barrel for the assembly of a lift-up compensator and a folding stock. *Jacques Barlerin*

The STAT defined the equipment that the army needed as specified by the army general staff. It also ensured the liaison among the operatives, the users of the equipment, and the organization tasked with supplying the necessary equipment: the General Delegation for Armament (DGA). The DGA then had the equipment made, or acquired it, as defined by the STAT.

In order to do that, the DGA appealed both to its private suppliers and groups of state industrial assets (e.g., weapons manufacturers). Immediately after the Liberation, an organization was created to manage and coordinate the actions of all the organizations of weapon research and production: the DEFA (weapons research and production unit).

The group of state industrial assets was regrouped in 1973 within the GIAT (Groupement Industriel de l'Armement Terrestre) group. After intense restructuring undertaken starting in 1991, GIAT took the name "GIAT Industries," then "NEXTER" in 2006.

Attempts at Modernizing the MAS-38 Submachine Gun

Even though replacement of the MAS-38 submachine gun had already been planned, production was continued up to 1950, so the French units were able to wait for the commissioning of the new MAT model 1949 submachine gun.

Various tests were carried out by the armies (and by the police, who also used many submachine guns of this model) in order to modernize the MAS-38:

- reduction of the protrusion of the firing pin to reduce the primer perforation
- trials with fluted barrels with a rate of twist of 270 to left, as on the 24-29 automatic rifle (the rate of twist of the barrel on the MAS-38 was initially 254 to the right)

Tests carried out in the applied ballistics research laboratory (LRBA) in February 1948 showed that the use of fluted barrel with rate

Parachutist of the colonial forces armed with a MAS-38 submachine gun. *GU*

The model 1935 and the first model 1938 submachine guns were fitted with a firing pin independent of the bolt (*A*), held in fixed position by a pin (*arrow*). The bolts of the model 38 made after the war (*B*) have a fixed firing pin. The rear part of the firing pin, serving as a guide for the spring on the first type of bolt, was replaced by a cylindrical part.

of twist of 270 to the left improved the accuracy of the model 38 submachine gun:

- grip plates in molded material, bipod and collimator for night firing, silencer, folding or retractable stock, etc.
- adaptation of the MAS-38 for 9 mm Parabellum ammunition
- trials of frames in pressed steel, light alloy, and even plastic material

However, because of the ammunition used, along with the difficulty of transforming the weapon into one with a folding stock (due to the presence of the recoil spring in the stock), the MAS model 38 no longer met the needs of the French forces. In addition, this weapon, made according to traditional machining methods, used prewar technology and no longer met the demands of rapid, large-scale, and low-cost manufacture of the postwar period.

Outside the modifications in the detail of certain parts, the MAS model 1938 continued to be produced in the same form as in 1940. Only the bolt was modified: the pinned firing pin on the interior of the moving bolt on the first models was replaced by a bolt with a fixed firing pin. An independent recoil spring guide was added at the rear of the bolt to

Before the general use of the MAT model 1949 and, to a far lesser extent, the model 1954 submachine guns by the police, solutions were sought to increase the compactness and the safety of use of the MAS-38 submachine gun. This photo shows a MAS model 38 modified by the technical services of the police headquarters: a safety lock was added to the rear of the grip, the trigger no longer flipped forward, a knob operated the fire mode selector on the left side of the receiver, the stock was replaced by a MAT-49 retractable stock, the magazine housing folded under the barrel, and the modified magazine had a hoop at its extremity, allowing it to be attached under the barrel. Many MAT-49 parts were used to carry out this transformation. *Private collection*

The future colonel Ponchardier before leaving Saigon on August 22, 1946, armed with his MAS-38 submachine gun.

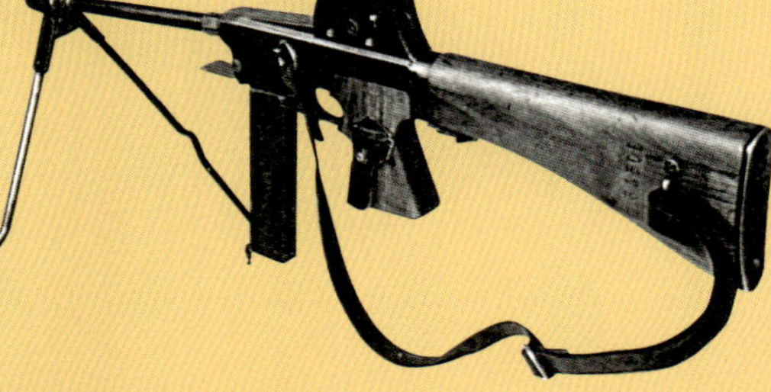

This example of a model 38 submachine gun fitted with a black collimator has a bipod, which apparently at the time did not increase the accuracy of firing in prone position. *CAA*

Prototype of the night-firing collimator (known as "black collimator") designed by squadron commander Bayrou of the 13th Artillery Regiment, stationed at Bourges. The rear sight eyepiece and the front sight were moved to the left of the weapon. At the center, the collimator has a lighting system, showing a horizontal, interrupted luminous line making the rear sight appear, and a vertical line making the front sight appear.
These two lines, with no distance limit, significantly improved the accuracy of fire in darkness. This device could also be adapted to the MAT model 49 submachine gun. This interesting project however was not developed. *CAA*

This photo of General de Montsabert, taken in Karlsruhe in April 1945, shows his driver and bodyguard carrying a Gnome and Rhône submachine gun. This is a rare document showing the weapon in service. *ECPAD*

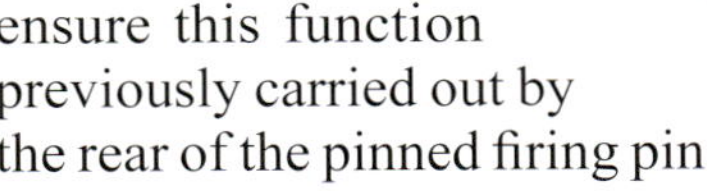

Gnome and Rhône submachine gun: copy of the Sten Mk. II made in Limoges at the Liberation

Safety indentation permitting the bolt to be blocked in a forward position

ensure this function previously carried out by the rear of the pinned firing pin.

It should be noted that the 38 submachine guns were delivered with carbine leather slings before and during the Second World War, and from the Liberation to May 1948 were delivered with slings in canvas (production corresponding to the end of series F and half of series G).

Gnome and Rhône Submachine Guns

The French army used a small number of Gnome and Rhône submarine guns, also called "STEN R5," during the last months of the Second World War and for the following ten years. This faithful copy of the British STEN Mk. II had been for a short time made at Limoges by one of the factories of the famous manufacturer of aircraft engines and motorcycles. It appears that this production was undertaken in the months after the liberation of Limoges on the initiative of the authorities of the Resistance who controlled the region at that time.

The plans were based on a Sten Mk. II made available to the design office of the factory by one of its senior staff, engaged in the Resistance. Even though the plans, still held in the armament archives, were strictly speaking those of a Sten Mk. II, the weapon that was produced differed from its British model by the presence of a wooden stock, a pistol grip placed under the barrel, a rotating flap under the receiver (allowing the bolt to be blocked in a forward position), a shorter barrel sleeve than on the Sten, and a slightly different body cap.

The number of Gnome and Rhône guns made was probably in the region of 8,000, as confirmed by an order for 8,000 subassemblies of spare parts with the Tulle factory for the Gnome and Rhône plant. The first weapons were received by the military authorities of the Resistance in August or September 1944. The R5 marking on these weapons certainly symbolizes the fifth mil-

Marking on the magazine housing of a Gnome and Rhône: the marking "R5" probably symbolizes the fifth military region of the Resistance, whose command ordered the manufacture of this weapon, from the liberation of Limoges in 1944. The Gnome and Rhône submachine gun subsequently remained known under the name "STEN R5" in the French army.

Sten Mk. II submachine gun. Large numbers were airdropped for the Resistance; Stens retrieved (at least partly) among the population by postwar army units were reused jointly with the weapons allocated by the Allies to the Free French forces and by the French army up to the end of the 1950s.

Sten Mk. 5 submachine gun. This weapon was used by the SAS of the France Libre. These weapons also were used in a small number by the Ponchardier commando, as well as by some units of the colonial paras at the beginning of the war in Indochina. This model is fitted with a "spike" bayonet and a forward pistol grip. This type of pistol grip seen in the photo was abandoned during production. On those submachine guns that had already been fitted with it, it was more often than not removed, since it tended to facilitate the accidental unscrewing of the barrel jacket when subjected to an impact from the left side.

itary region of the Resistance, the region where Limoges was situated. The last deliveries of the Gnome and Rhône took place in November 1945. Thereafter, the central power, which had reestablished its authority over all the territory, undertook to reserve the manufacture of war weapons to state-owned factories.

The weapons are numbered from 001 to 8000 on the magazine housing, which also has the marking "R5" and another series of figures, the significance of which remains unclear (perhaps the month of assembly of the weapon?). Before the definitive cessation of production, a last series of 100 "R5 submachine guns" was produced for the DGER (Directorate General for Studies and Research), the ancestor of the DGSE.

There is at least one photo of the period that shows this weapon in service during the German campaign in 1945. In 1952, the Sten R5 was still on the list of weapons in service in the French army.

The 1952 edition of the "Units Supply Chart" of the French army still mentions the Gnome and Rhône submachine guns in its inventory. It is likely that the "R5 submachine guns" partly disappeared during the war in Indochina, and that the specimens remaining in France were subsequently destroyed along with the rest of the Sten still held in French military establishments.

FOREIGN SUBMACHINE GUNS OF THE FRENCH ARMY

In 1944, the French army was being reconstituted and found itself equipped with a large variety of submachine guns: the MAS-38 and Thompson 1921, which equipped the armistice army up to 1942, represented a tiny part of the weaponry. The main submachine guns in service at the time of the Liberation were of British, German, or American origin.

British Submachine Guns

The great majority were Sten, and a lesser number were Lanchester.

The Sten Submachine Gun

The majority were the Mk. II model, and they came from various sources:

- units of Free French forces armed by the British
- integrated Resistance units, with their weapons coming from Allied airdrops for the French army of the liberation
- weapons retrieved at the end of the war, when the Resistance was disarmed
- incidentally some Stens, captured from the Resistance and reused by the German forces or units of the French militia joined up with the first ones

The Sten Mk. II was widely used by partisan units fighting alongside French forces.

In the French military archives, plans dated 1946 established by the LRBA (Applied Ballistics Research Laboratory) were found of various models of bolt handle allowing the bolt to be immobilized in a forward

Sten transformed by the addition of a cooling jacket at the base of the barrel, and a pistol grip under that. This trial, carried out by the research section of the MAS, was discontinued. *Jacques Barlerin*

The Lanchester Mk. I submachine gun was above all used by the navy, but also by the Ministry of the Interior.

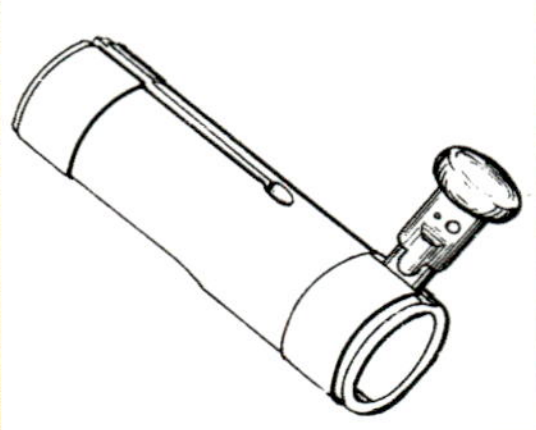

The bolt on some Lanchester submachine guns was modified by the addition of a type MP 40 bolt handle, allowing the bolt to be blocked in a forward position. The initial-type bolt reference is 61, and that of modified bolts is 61A.

position, following the principle of the British Mk. 5 bolt handle.

There was no trace, however, of a directive ordering the transformation of the weapon, in service in the armies, for the use of this part. The introduction of these new bolt handles was systematically carried out on Stens used by the Ministry of the Interior.

Stens with silencers, dating from the Second World War, were also used by commando units. Silencers for Sten submachine guns were researched in 1952 and 1953 by the ETVS (silencer: ETVS types 522 and 531). But these accessories, despite their effectiveness when used with subsonic cartridges (the study and manufacture of which were launched in France for the occasion), did not go on to be developed. They arrived too late, at the time when the war in Indochina was coming to an end and when the Sten submachine guns were starting to be replaced systematically by the MAT model 1949 within French units.

Subsequently, the Stens remaining in the depots of the French army were gradually destroyed, with the exception of the ones on which the mechanism was completely welded and that were issued to commando training centers as a substitution for the MAT-49 to carry out different combat tasks.

We have found no trace of the use of the Sten Mk. III by the French army; the Sten Mk. 5, on the other hand, was issued to French units in the Far East trained by the SOE (Special Operation Executive), who were among the first to intervene in Indochina in 1946.

The Lanchester Submachine Gun

A small number of Lanchesters were parachuted in the Southwest of France, but it seems that very few of them were recovered at the end of the war. The Lanchester submachine gun, however, was the weapon on board certain Royal Navy vessels used by Free French naval forces or sold to the navy

Thompson 1928 A1. Mechanically identical to the 1928 model, this version was most commonly equipped with a horizontal handguard. This model was used by many units of the French army up to the 1960s.

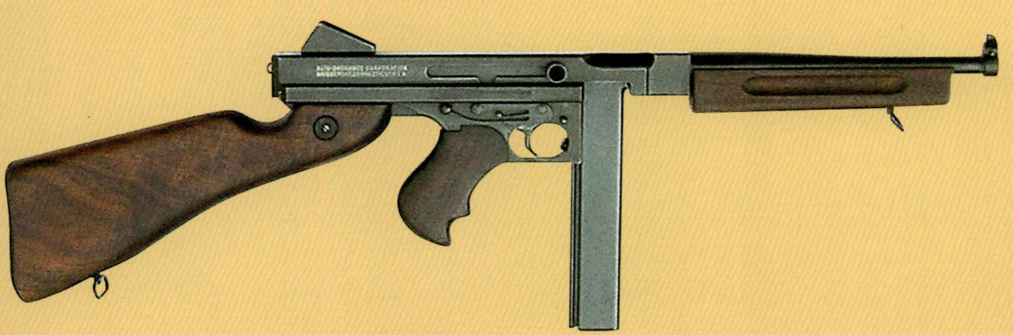

Thompson M1A1 submachine gun. After the American landings in North Africa in 1942, large quantities of the Thompson model 1928 A1, M1, and M1A1 were allocated to French forces in Africa united with the Free French, numerous units of which were totally reequipped by the Americans. After the Second World War, new batches of different models of Thompson were delivered to France by way of American military aid.

Commando of the French navy armed with a Thompson submachine gun. *DR*

US M3 submachine gun in .45 caliber. From the same origin as the Thompson, the M3 and M3A1 were used in fewer numbers than the Thompson.

after the war with all their onboard equipment. These Lanchesters remained in service for a time in the navy. Some of them finished their working life on board launches of the Paris river police or within the CRS (Compagnies Républicaines de Sécurité).

American Submachine Guns

Thompson Submachine Gun

The French army had tested various models of Thompson from the 1920s and ordered 3,000 Thompson 1921s from the United States in 1940.

During the occupation, small quantities of Thompson model 1928 A1, M1, and M1A1 submachine guns were parachuted to the Resistance. In parallel, some North African units reequipped with American weapons received significant quantities of 1928 A1s, M1s, and M1A1s, which were used during the French and German campaigns. The French marine commandos trained in Great Britain were also issued with Thompson 1928s and 1928 A1s.

At the end of the Second World War these weapons continued to be used by all the French army, which subsequently received new deliveries by way of American military aid. The various models of Thompson submachine guns remained in service up to the end of the war in Algeria and were then replaced by the model 1949 submachine gun.

M3 and M3A1 Submachine Guns

Of the same origin as the Thompson submachine guns, the M3 and M3A1 were also used by the French army, though in substantially fewer numbers than the Thompson.

The Thompson, M3, and M3A1 submachine guns fired the .45-caliber ACP cartridge, known as "11.43 mm" in the French army. To complete the batches supplied by the United States in the 1950s in the context of military aid, this ammunition was also made for the French army by the SFM (French Ammunitions Company).

German Submachine Guns

A significant number of MP 40s were recovered from the occupier. These weapons seem to have been used essentially in Indochina by Marine commandos, but also in units of the Foreign Legion.

It must be remembered that a small number of German MP 44 assault rifles were also used at the beginning of the war in Indochina by several units such as the 13th Half Brigade of the Foreign Legion (13th DBLE).

2nd Stick of the 1st Commando / GC3 (3rd Company Commando Group) of the 6th BCCP (Bataillon colonial de chasseurs parachutistes), photographed in Annam, Tourane, in 1950. The three men in the foreground are armed with the MP 40. The locally made cartridge belt on the para on the right can clearly be seen. The second man on the left, armed with a pistol, was the shooter of the 50 mm grenade launcher of the group. *DR*

German MP 40 submachine gun; many weapons of this type captured from German troops were put back into service in French units.

CHAPTER 9

RESEARCH CARRIED OUT IN FRANCE FOR THE PURPOSE OF INTRODUCING A NEW NATIONAL SUBMACHINE GUN

One of the first versions of the Hotchkiss submachine gun: the model 011, closely based on the Sten. *Jacques Barlerin*

Hotchkiss model 304 submachine gun. *Jacques Barlerin*

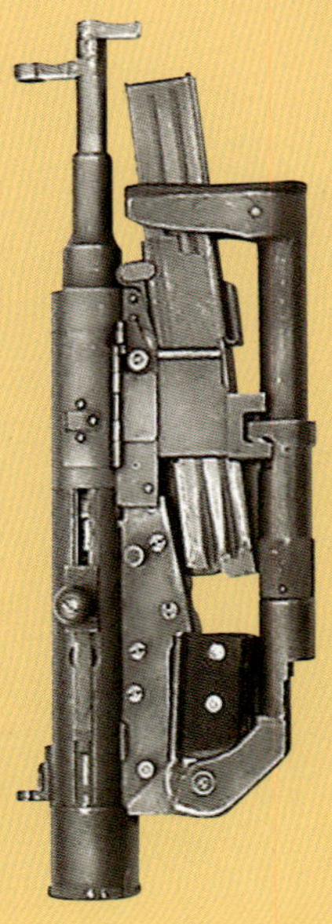

The 010 model folded. In order to reduce the volume of the weapon, the barrel can be pushed inside the receiver.

Specifications of the characteristics of a future automatic pistol and a future submachine gun for the French army, written in October 1946 by the technical section of the army, specified, "With regard to the submachine gun, it must be in 9 mm Parabellum caliber and be equipped with a fold-back or retractable stock and a magazine folding under the barrel. It must be able to be manufactured by methods necessitating a minimum of machining and a maximum of parts and components made by pressing and welding."

Several private enterprises presented the French army with prototypes answering to these specifications to various degrees.

In addition, the three national weapons manufacturers—Saint-Etienne (MAS), Tulle (MAT), and Châtellerault (MAC)—also entered the running and presented numerous prototypes to the French army.

WEAPONS PROPOSED BY FRENCH PRIVATE INDUSTRY

Hotchkiss Semiautomatic Carbine

The Hotchkiss company acquired a great reputation starting in the last years of the nineteenth century by making revolving cannon and air-cooled machine guns according to the Odkolek patent. Between the wars, Hotchkiss expanded its activities to the production of heavy machine guns and vehicles, and after its merger with the Brandt firm, Hotchkiss made mortars and artillery ammunition.

Just after the Second World War, Hotchkiss developed a line of three submachine guns with selective fire, called "semiautomatic carbine." The only model to have known a certain circulation was the Hotchkiss model 010 semiautomatic carbine (also known as

Hotchkiss semiautomatic carbine model 010, also known as "CHM2" or "universal type"

Three submachine guns typical of the beginning of the war in Indochina: MP 40, of German origin, a weapon that was widely used by the French expeditionary force in the Far East; Hotchkiss semiautomatic carbine; and early-made first-type MAT-49. ***Marc de Fromont***

Variant of the model 010 semiautomatic carbine with folding front stock, presented by the Hotchkiss company during the ETVS tests

the "Universal type" or "CHM2"), which has the original peculiarity of being fitted not only with a folding stock and magazine, but also with a barrel able to be partially retracted in the receiver for transport purposes.

The Hotchkiss submachine gun did not interest the French army, since it was too complex, not user-friendly, and too sensitive to foreign bodies entering the mechanism.

In the postwar world, where Second World War surplus was abundant and accessible at a low price, the Hotchkiss did not attract buyers. The remainder of the 700 specimens made by the firm at Saint Denis were exported to Laos, Cambodia, or Morocco. A photo showing a soldier of the Vietnamese army armed with a Hotchkiss submachine gun presumably means some were also sent to Indochina.

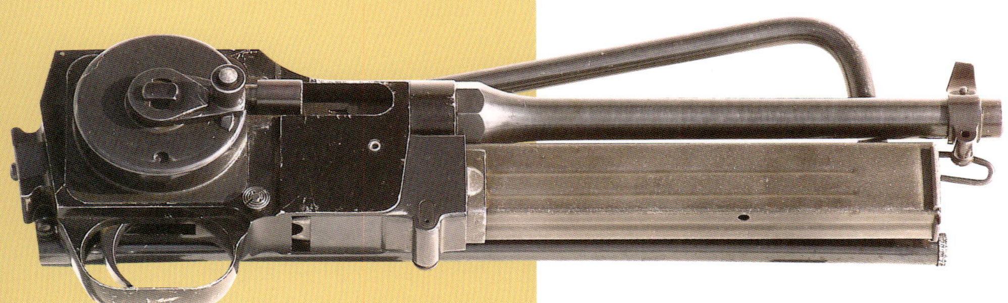

The MGD submachine gun with metal folding stock. *Collection WTD 91, Meppen*

MGD Submachine Gun

This highly original weapon was developed by the Merlin-Guérin-Dauphiné company, which was initially dedicated to the production of electrical devices. Immediately after the Second World War, this company patented numerous mechanisms destined for the armaments industry, among which were ingenious and original systems for various weapons designed by a company engineer, Louis Debuilt: a machine gun, an automatic pistol, a semiautomatic rifle, and a submachine gun. Even though none of these inventions were mass-produced, the French army tested several versions of the MGD submachine gun with a great deal of interest. The weapon was very compact and in particular had a very short receiver obtained by the use of a short, nonfixed bolt, solid with a rotating piece itself linked to a spiral recoil spring. The hinged magazine housing meant the magazine could be folded under the barrel.

The weapon was tested starting in early 1947; the last test reports found in the weapons archives deal with a prototype on which the spiral recoil spring was replaced by a cylinder of compressed air, dating to 1955. Despite the fact it was easy to handle and its accuracy was highly satisfactory, it presented many firing incidents and proved to be sensitive to the entry of foreign bodies. In addition, the rates of fire measured during the tests were irregular, probably due to its recoil spring; its very complex design contributed to its rejection in favor of other prototypes better adapted to an economical, large-scale production.

The West German firm Erma-Werke planned to manufacture the MGD submachine gun under license but eventually abandoned the project, as much due to fragility of the mechanism as the very high production costs that would be involved.

Delacre Machine Pistol

The history of this private inventor remains relatively unknown even today. Henri Delacre had registered the patent for a small-size machine pistol in 1936 that could be fired with a single hand, and with a semicircular latch at the rear of the receiver that fit around the shooter's forearm. Delacre was also responsible for using a very powerful 6.35 mm ammunition: the 6.35 Delacre (6.35 × 27.5), which had been studied by the French Ammunitions Company and was not followed up on.

Immediately after the Second World War, Henri Delacre proposed a surprising machine pistol of his own design to the French army.

MGD with folding wooden stock. Some prototypes of this model were made in 7.65 mm Long and 9 mm Parabellum. Some with a frame in duralumin were also made. *CAA*

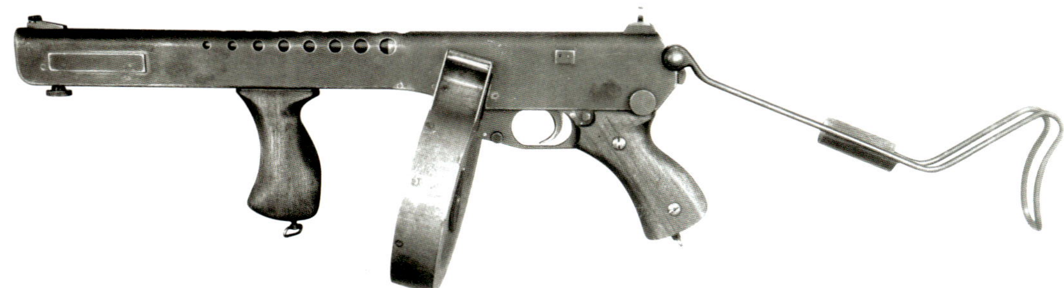

Prototype of the Delacre machine pistol in 9 mm caliber Steyr tested by ETVS just after the Second World War. The total length of the weapon was 800 mm with the stock extended, and 540 mm with folded stock. Its barrel measured 317 mm, and its empty weight was 3.29 kg. A loaded magazine weighed 1.16 kg; therefore, the total weight of an operational weapon was 4.45 kg. *Jacques Barlerin*

The report made by the ETVS on June 11, 1948, states that for more than one year, the experimentation process consisted only of a development of the weapon, the flaws reported to Satory having been pointed out to the maker, who tried to remedy the situation each time! It was not before December 1947 that more-thorough experimentation was possible (it can be concluded therefore that the first prototype of the Delacre machine pistol was presented to the ETVS in 1946).

Curiously, this weapon is chambered for the 9 mm Steyr ammunition, which was presented in the ETVS report as "9 mm Star." The 9 mm Steyr (9 × 23 mm) was very close to the 9 mm Largo or Bergmann Bayard, which was regulation in Spain and tested by the ETVS before the war in various "bursting" Spanish pistols or machine pistols, so the confusion in terms is easily explained.

The choice of this unusual ammunition was probably a response to the desire to obtain a "universal" weapon that was more powerful than the standard MP, without passing to the stage of intermediate ammunition developed at the end of the war for assault rifles such as the German MP 44. The weapon, fed by a fifty-round drum magazine, was operated by a long recoil of the barrel, ignition was ensured by a hammer, and firing was single shot and "full auto." The Delacre machine pistol was fitted with a case detachment device facilitating extraction. The ETVS judged the weapon too complicated and operationally unreliable, particularly since in the inventor's mind the magazine was to be used only once and was made with parts that lacked thickness and robustness.

Drum magazine, Delacre machine pistol. This single-use magazine was ahead of its time in terms of its concept. However, its lightweight construction rendered it too fragile. *Jacques Barlerin*

Gévarm MP

This weapon was developed during the mid-1950s by the "La Gévarmoise" company, a subsidiary of the Gévelot firm. Two versions were tested by the ETVS: the model D4, with a retractable metal stock, and the D3, with fixed wooden stock. The Gévelot submachine gun did not interest the French army, and only the model with the metal retractable stock was produced in a small number in two versions: one with a safety lever forward of the grip, and the other without a safety lever. A small quantity of Gévelot submachine guns were issued to the CRS (Compagnies Républicaines de Sécurité) and the police prefecture in Paris.

Others were exported to Africa. On the occasion of the closure of "La Gévarmoise," a small batch of unsold Gévelot D4s remained in the factory stockroom. A number of these weapons was bought by a cinema props department to be transformed to fire blanks; others were sent to be deactivated at the Official Bench for the Testing of Arms and Ammunition at Saint-Étienne before being sold on to collectors.

The D4 submachine gun with unfolded stock. Gévarm was a subsidiary of the Gévelot company. It also made a semiautomatic carbine in .22 LR caliber. The submachine guns were made at the "La Gévarmoise" factory at St. Germain Laval. A wooden stock version and a retractable metal stock version (serial numbers 353 043 and 353 020, respectively) were tested by the ETVS on February 19, 1957. *Jacques Barlerin*

The model D3, version with a wooden stock from Gévarm, presented to the ETVS in 1955

PROTOTYPES PROPOSED BY STATE MANUFACTURERS

MAC Prototypes

After France's defeat in 1940, the Arms Manufacturer of Châtellerault (MAC), finding itself in occupied territory, was placed under the control of a German official. While the occupier was taking over a part of the machine tools in order to transfer them to Germany, the MAC was authorized to take on some production activities purely for the civilian market.

Gradually the personnel working under German control found themselves compelled to make weapons parts and bayonets for the K.98k carbine.

Acts of sabotage and the refusal to carry out mandatory work led to the deportation of thirty-six members of staff and the execution of nine others. Before their departure at the end of August 1944, the occupiers took numerous machines, plans, and auditors back to Germany.

Despite everything, MAC was able to restart production of the FM 24-29 starting in November 1944.

In 1947, in order to conserve sufficient activity and maintain its personnel, as the MAS and the MAT had done, the MAC made hunting rifles with a juxtaposed barrel, eagerly requested by a population that wanted to replace the weapons that had been confiscated or destroyed during the war.

While exploring several prototypes for carbines (model 1948 and 1949 carbines) based on the American M1, the factory at Châtellerault set up a study of a series of prototypes in 9 mm Parabellum caliber, destined to compete for the choice of a new submachine gun in the French army.

The first two were finished in 1947: these were models 1947-1 and 1947-2, which in certain aspects were reminiscent of the ETVS submachine gun, of which fifty preproduction models were made by MAC in 1939. The models 1947-1 and 1947-2 had the following features: a moving bolt but with delayed opening, a "rat trap" type of recoil spring housed in the trigger guard, and a folding stock and magazine, firing only in short bursts.

- The 1947-1 model has, as does the ETVS, a magazine under the barrel and a metal, laterally folding stock on the left side of the weapon. Its strengthened barrel was surrounded by a cylindrical jacket reminiscent of the Sten Mk. I and Mk. III. The bolt handle, placed on the right side of the weapon, is based on that of the USM 3 submachine gun.

- The 1947-2 model is fitted with a cleared barrel and a stock folding over the receiver. Curiously, the bolt handle on this submachine gun is placed under the trigger guard. The movement of the bolt handle leads to the recoil spring lever rotating to the rear, causing the recoil of the mobile group.

MAC model 1947-1 with strengthened barrel and folded stock. The weapon is fitted with a bolt handle on the right side. This view shows the bolt with its delayed-opening system and the cocking-handle type of bolt handle based on that of the USM3 submachine gun.

MAC model 47-2 submachine gun, which differs from the 47-1 model by its stock that folds over the weapon, its cleared barrel, and the bolt handle placed under the trigger guard

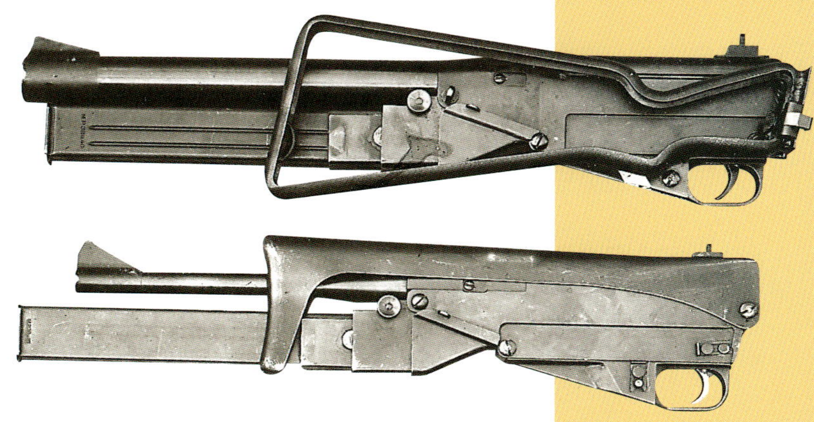

Comparison between the MAC model 47-1 and 47-2. ***CAA***

Shortly after, the light modified 1947-2 and 1948 LS (LS standing for "light, simplified") models appeared with a Sten-type bolt and a tubular receiver. The weapon had a folding magazine under the barrel, a sliding metal wire shoulder stock (US M3 type), and a safety lever at the rear of the grip. The barrel was encased in a perforated jacket. The rear part of the receiver could be disassembled in the same way as the Italian FNAB 43 submachine gun in order to access the moving bolt.

In the end, none of the MAC prototypes were used. However, after the adoption of the MAT model 49 submachine gun by the French army, the MAC was tasked with the manufacture of a part of these weapons.

MAS Prototypes

Between the wars, the MAS played a vital role in the research on submachine guns undertaken in France. As has been seen previously, the "research section" of the MAS produced a small modified series of the SE MAS-35 submachine gun in the first half of 1940 while waiting for the assembly lines to start up.

After the French defeat in 1940, the MAS, situated in an unoccupied zone until 1942, made MAS-38 submachine guns for the armistice army.

After the invasion of the free zone, production was placed under the supervision of German inspectors seconded from the Mauser factories. Toward the end of the war, the occupier set up the production of receivers for G.43 semiautomatic rifles. Starting with the liberation of Saint Etienne in the summer of 1944, the MAS quickly recommenced the production of French-type weapons: MAS-36 rifles, MAS-38 submachine guns, and then a modernized version of the MAS-40 semiautomatic rifle: the MAS-44.

The army general staff had chosen the 9 mm Parabellum ammunition as the regulation cartridge for the future submachine guns and automatic pistols of the French army in 1946. As a result, trials were undertaken to adapt the 38 submachine gun to the 9 mm Parabellum ammunition, as was seen previously in the chapter dedicated to French submachine guns of the Liberation. From 1944 to 1947 the research section of the MAS also led diverse experiments on the production of the model 38 submachine gun with a frame in pressed metal. Production tests of frames in light alloy and in plastic material were also carried out. These experiments did not result in any corresponding commissioning, but they permitted the research units to acquire the mastery of new production techniques that had been developed by the Germans, Soviets, and Americans during the conflict. These various attempts, however, did not go beyond the prototype stage.

MAS Model 1948 Submachine Guns

In parallel, the MAS devoted important funding to establish a new family of individual automatic weapons, characterized by a bolt with delayed opening and a receiver in pressed metal, based on that of the German Sturmgewehr.

SE MAS-48 type A submachine gun bearing the serial number 1. ***Jacques Barlerin***

The method of construction and operation chosen by the MAS presented the significant advantage of being just as suitable for the manufacture of a 9 mm caliber submachine gun as a rifle firing an intermediate-power cartridge, between that of the rifle and that of the pistol or the .30-caliber cartridge of the American M1 carbine.

The "assault rifle" versions, designed to fire an intermediate cartridge, were christened "carbines." They had longer barrels, fixed wooden stocks, and trigger mechanisms allowing for selective fire. Some were designed to fire intermediate ammunition then studied by the technical services of the French army, and some others were chambered for the 7.63 × 33 mm cartridge of the US .30-caliber carbine.

A series of tests were carried out with prototype MAS submachine guns with anatomical shoulder stocks of various shapes designed to improve the support of the shooter's forearm. These stocks were developed in association with the noncommissioned officers' schools at Saint Maixent and Strasbourg. ***Jacques Barlerin***

MAS model 48 C4 submachine gun. The model C4 was an improved version of the model C3, with a magazine that folded back and unfolded in a single movement (on the C1, C2, and C3 models, it was necessary to push the magazine toward the rear after having folded it under the barrel in order to hold it in position). The grip is longer than on previous models and much more comfortable for the shooter.

A submachine gun of this design in 9 mm caliber Parabellum was the MAS-48, several versions of which were presented at the tests where the choice of the new submachine gun of the French army had to be made.

The versions of the MAS model 1948 submachine gun (C1, C2, C3, C4, and D) were distinctive by the length of their barrel, the type of stock used (fixed in wood folding laterally, metal wire folding under the weapon), and various improvements in detail (reinforcement of the magazine housing, fixing of the sling, etc.).

At the conclusion of a first series of tests, which resulted in the elimination of prototypes of submachine guns presented by the MAC, a small series of MAS-48 type C and MAT model 1948 light submachine guns were made in order to be trialed in combat units, in particular within the units of the French Expeditionary Corps in the Far East.

MAS-49 submachine gun in 11.43 mm caliber, bearing the serial number 4. ***Jacques Barlerin***

MAS Model 1949 Submachine Gun

This improved version of the model 1948 was equipped with a folding magazine housing, a retractable stock in metal wire, and a safety lever at the rear of the grip. Apart from the version in 9 mm caliber Parabellum, some examples in .45 ACP caliber were also researched.

Despite its great qualities, the weapon remained more complex and fragile, however, than the prototype proposed by the MAT. Since the general staff had abandoned the adoption of an assault rifle in the short term, the homogeneity of design between the submachine guns and the semiautomatic carbines of the MAS was no longer considered an argument likely to lead to the acceptance of the relative complexity of the submachine gun and left the field open for the prototype proposed by the MAT.

MAT Prototypes

After the invasion of the free zone by the German forces in 1942, the MAT was placed under the control of a team of German technicians from the firm DWM. Some equipment was seized by the occupier and taken to Germany, while the factory personnel were forced to work under German control. The MAT produced, among other things, certain parts for the MG 131 heavy aviation machine gun.

When the Resistance besieged the town of Tulle in June 1944, it was quickly recaptured by the Waffen-SS division "Das Reich," and ninety-nine of its inhabitants were arrested by the Germans and hanged; this figure included thirty-three employees of the weapons factory. Shortly thereafter, the enemy decided to move all the equipment able to be dismantled from the MAT to Germany and to deport 400 workers from the factory. The convoy transporting men and material finally stopped at Epernay. Part of the material was stored in the champagne cellars of the Maison Mercier, while the remainder (around 350 machines) continued on to Germany. The arrival of American forces prevented the MAT employees and the stored material from completing their journey.

Immediately after the Liberation, in a France that had been devastated by war, the reequipping of the MAT could take place only progressively. Thanks to some machines supplied by the Châtellerault factory, the MAT could start minimal activity by producing barrels for 24-29 light machine guns, parts for the model 31 machine gun, and the 25 mm antiaircraft cannon, along with a collection of 8,000 parts for the Gnome and Rhône submachine gun, destined for the Limoges factory, which in 1944 had undertaken the production of this French copy of the Sten.

After the renovation of the machines coming from Epernay, the MAT carried out various orders for the civilian market, as did other national weapons manufacturers, for tooling, hunting rifles, and the like in order to conserve its potential for production in the event of future military orders.

First prototype of the MAT submachine gun made in Tulle starting in 1945. Unfortunately, this prototype was destroyed in Satory in 2004. *Jacques Barlerin*

MAT Model 1947 No. 1 and No. 2 Submachine Guns

After the Liberation, some MAT technicians made a prototype of a compact submachine gun. Only a photograph of a prototype dating from 1945 has survived the destruction of this weapon in 2004 in Satory.

The general design of this weapon was based on the MAT model 1947 no. 1. The same year, another prototype appeared that was quite different in terms of its appearance and its operational principles: the MAT model 1947 no. 2, operating with a fixed bolt. Unlocking was carried out by lowering the barrel by means of a connecting rod as on the Colt or Browning pistol systems.

The MAT model 1947 no. 2 had the interesting peculiarity of a bolt surrounding the barrel. This design, which resulted in the weapons being quite short and very stable during full-auto fire, was later present on Czech submachine guns of the VZ 25 series and on the Israeli UZI, which in turn went on to inspire many other submachine guns.

Another particularity of this weapon: when the magazine was folded under the barrel, a sealing flap solid with the magazine housing concealed the ejection port.

The weapon had a safety, but action on the trigger was prevented while the magazine was not in place. On the other hand, the weapon did not have a safety in case of inertia recoil of the bolt (occurring if the weapon fell accidentally, for example).

As a result of the tests, the commission noted that the magazine safety, based on those on certain automatic pistols, could not be justified on a submachine gun, where the firing stop is produced by the bolt locking rearward on the sear.

However, this type of safety could cause accidents, since it necessitated putting a magazine in place in order to disarm the weapon, with catastrophic results if the user inadvertently took a loaded magazine.

1947 MAT no. 1 of 1947. *Jacques Barlerin*

Nonmodified MAT no. 2 prototype in position for transport. *CAA*

Prototype of the modified MAT no. 2. On this version the magazine safety, considered unnecessary and a source of accidents, was removed. The stock was slightly lengthened to be used as a muzzle cap and prevent the loss of the magazine when the weapon was folded for transport. *CAA*

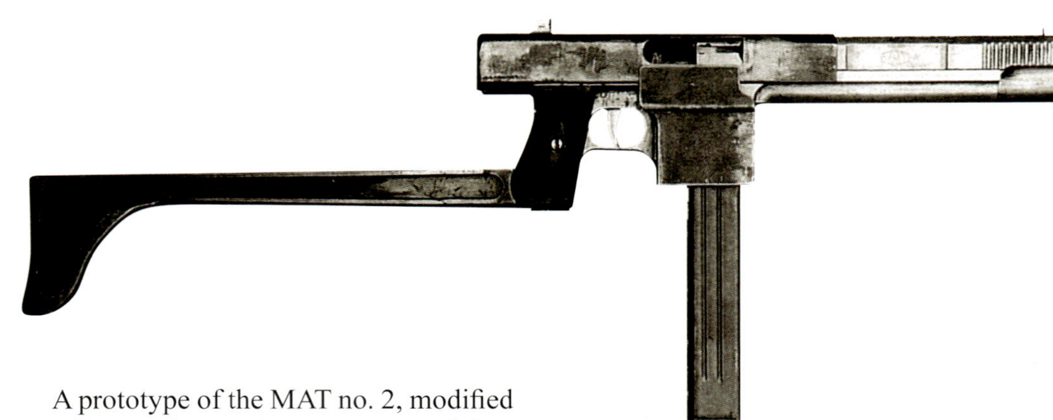

The modified MAT no. 2 still had no safety preventing the inertia recoil of the bolt. Conversely, it is fitted with a hook-shaped part, solid with the trigger, which prevents the bolt from taking a cartridge in the magazine if it moves forward while the shooter is not pressing the trigger. *CAA*

A prototype of the MAT no. 2, modified according to the requirements, reacted in a satisfactory manner during a new series of tests. The vibrations during full-auto fire were greatly reduced compared with the initial model, thanks to the reduction of the depth of the chamber, and consequently the bolt abutted the case and no longer the rear section of the barrel during closing.

MAT Model 1948

The MAT nos. 1 and 2 seem to have been designed to offer their users a very compact self-defense weapon, which could be transported folded and be implemented in a crisis situation. The MAT model 1947 no. 2 in particular resembles an oversized automatic pistol.

With the MAT model 1948, the Tulle factory appeared to have opted for an offensive weapon, intended to be carried on a permanent ready-for-use basis, but with a folding stock and magazine making it more compact during transport.

From a mechanical point of view, this new line of prototypes clearly showed a greater simplicity: the bolt on these submachine guns was a single breech block devoid of any system to delay the opening. According to the terms of the manufacturer, "The study was conducted with the concern of making a weapon that could be mass-produced for a very low price, without sacrificing operational safety and efficiency in any way. We have used pressed sheet metal as much as possible, and, combined with electrical welding, this manufacturing process was fast and economical. A small number of easily produced machined parts are used."

Two versions of this weapon were proposed initially:

- The model known as "heavy," with a 370 mm long barrel and a fixed stock and the light model with a 230 mm barrel and a folding stock. The heavy model weighed 3.5 kg as opposed to 3.2 kg for the ordinary model.

- A model known as "modified" appeared shortly after. It is characterized by the following:

  1. The locking of the stock in a folded position on the left side of the weapon on the actuator knob. Another version with metal stock folding under the weapon was also presented by the MAT.

  2. the implementation of a system ensuring the magazine locking to the barrel when it is in a folded position

  3. an ejector solid with the trigger guard and not the receiver

  4. a safety device rendering firing possible only when the shooter holds the grip fully, which compresses a safety lever

MAT model 48 light submachine gun. This version has a metal hoop placed around the front sight to hook the sling, which passes to the rear in a circular orifice pierced in the stock, as on the German MP 38 and the French Gnome and Rhône submachine guns (some collections of parts were produced four years earlier by the MAT). With the 1948 model, the MAT abandoned the concept of an ultracompact submachine gun, previously studied with the MAT no. 1 and no. 2.

CHAPTER 10

THE 9 MM MAT MODEL 49 SUBMACHINE GUN

A series of tests beginning in October 1948 put the light MAT, MAS C4, and MAC 1947-2 prototypes in competition with one another. The MAC model, lacking robustness and showing a poor resistance to mud and dust, was eliminated.

In terms of ease of handling, the MAC 1947-2 submachine gun was, however, considered to be of interest, and the organization of the grip of the MAC 1948-2 prototype with a grip safety at the rear interested the testing commission.

Concerning the two other weapons, the report from the army technical section (STA) tasked with the experimentation process specified: "The light MAT does not yet present the ease of handling qualities vital for a crisis weapon, but the STA notes that several modifications of detail could render it excellent, whereas if the MAS C4 presented the ease of handling qualities required at this stage, it hardly seemed improvable."

The STA recommended the manufacture of 250 specimens of an improved version of each of the two prototypes retained after an initial series of tests:

- the light MAT
- the MAS C4

The improved version of the MAT light model 1948 incorporated various features considered of interest on the MAC prototypes, including the protective jacket around the barrel, designed to avoid the shooter's firing hand from receiving burns during sustained fire, and the pistol grip with a grip safety at its rear part.

The first versions of the MAT model 1949 submachine gun, called "short frame," are identifiable by the cube-shaped part at the rear of the frame, where the rear part of the receiver is housed.

French combatant in Indochina, armed with a first-type 49, recognizable by its cube-shaped frame. ***ECPAD***

Indochina 1954: the MAT-49 can be seen in all combat situations.

Cover from the June–July 1953 edition of *Indochine* magazine, showing a recruit of the new Vietnamese army equipped with a MAT-49 during training

These preproduction weapons were trialed in combat units, particularly operational units and schools.

The improved version of the light MAT submachine gun, with a grip with rear grip safety, a sliding metal wire shoulder stock, a simplified folding magazine housing, and a wider trigger guard, was a clear preference when compared to the improved MAS C4 for the testing units. This preference was shared by the STA, who appreciated the hardiness of the weapon and the possibility of producing it at a low price.

The MAT submachine gun was officially adopted on May 20, 1949, under the name "MAT model 1949 9m/m submachine gun," a name that was often abbreviated to "MAT-49" or "PM 49."

The model 49 was the result of teamwork led with passion by a group of MAT engineers, technicians, and workers operating under the direction of engineer in chief Monteil, the head of research at MAT. The use of parts made by stamping, which were then shaped and welded, allowed the time and cost of production to be drastically reduced, and these measures were inaugurated in France during the First World War for the manufacture of the French model 1915 light machine gun (CSRG 1915, also known as "Chauchat"). But the low need for weapons after the First World War meant these intensive methods of production had disappeared in favor of carefully machined parts. In the aftermath of the Second World War, it was necessary to catch up with the progress made in this domain by the other warring parties during the conflict.

Throughout the entire development of the weapon, head engineer Monteil benefited from the assistance of the head of production, engineer in chief Delamaire, another enthusiastic supporter of small-caliber weapons.

A first order of 5,000 specimens was communicated to the MAT from the early months of October 1949. The production of the weapon, which was carried out extremely quickly, was no small feat for the production personnel in the difficult postwar period.

In its initial version, the model 1949 was made up of eighty-seven parts; thirty-five of these were main parts, nineteen of which were produced by stamping.

The first fifty submachine guns were finished in January 1950.

STANDARD VERSIONS OF THE MAT-49 SUBMACHINE GUN

Throughout its production, the MAT model 49 benefited from certain modifications of detail of each of its parts, with the objective of these two goals:

- improving the solidity and the operational safety of the weapon
- reducing both the cost and time of production

Therefore, many successive versions of the same part can exist, differing from each other only by minor, sometimes barely visible, details. To create an exhaustive inventory of each one of these developments, in general very minor, would result in a tedious catalog outside the scope of the spirit of this special edition.

First-type model 49 submachine gun with a box of American-made Winchester ammunition and canvas items displayed on a Legionnaire's uniform from the Indochina war. *Marc de Fromont*

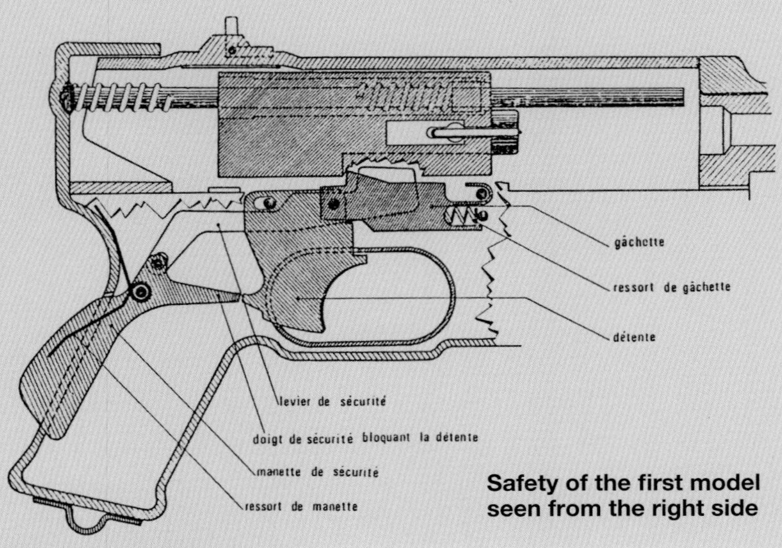

Safety of the first model seen from the right side

Trigger of a 49 submachine gun without the five-pointed star, fitted with the first-model safety. ***Frédéric Devolte***

We have chosen therefore to restrict this exposé to the presentation of the three principal production versions of the standard MAT-49, with the presentation of some variants produced in a reduced number in the following chapter.

MAT-49 First-Type Submachine Guns

This version, also known as the "MAT-49 with short frame" or "series A," had the following characteristics:

- frame and a recoil spring guide measuring 5 mm less than the corresponding parts on subsequent models
- cube-shaped frame housing
- Safety lever with a more rounded shape than on later models; this lever was made in Bakelite, pressed steel, or nylon.
- These weapons are fitted with a first-model trigger mechanism, identifiable by the absence of the five-pointed star marking on the trigger.

In addition, in their initial version these submachine guns were without the sling fixing hoop at the bottom of the pistol grip and the brace at the rear part of the stock serving as a butt plate. Later, a metal trigger guard would be welded onto the base of the grip on the MAT-49. This part was used by parachutists to facilitate the attachment of their weapon under their reserve parachute before they jumped.

The finish of these first-type MAT-49 weapons was generally a little more basic than on second-type models, and some minor differences in checkering on the catches can also be observed.

On the MAT-1949 model of the first type, the assembly of the rear of the frame to the receiver was ensured by a cube-shaped part welded at the rear of the frame.

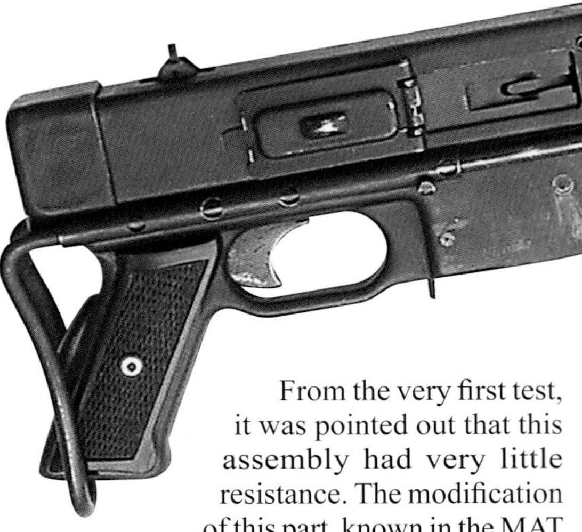

MAT-49 with short frame with a completely retracted stock and magazine folded for transport. In this configuration, the submachine gun could easily be positioned in a reserve parachute.

From the very first test, it was pointed out that this assembly had very little resistance. The modification of this part, known in the MAT workshops as "the cube," posed difficult problems at the time of manufacture due to the lack of inspections of the way certain pressed pieces were made.

Considering the urgency in supplying a modern submachine gun to the units engaged in the Far East, it was decided to start manufacture of the MAT-49 model submachine gun immediately in its first-type version; even though the design of frames with the "cube" could affect the life of the weapon, it was nonetheless not incompatible with its operational use.

Second-Type MAT-49

While the "first type" MAT-49 model was coming off the production lines and being delivered to combat units, the MAT worked to improve the assembly of the receiver to the frame so as to have a more durable weapon. After consultation between two German technicians, both specialists in pressing and employed in Alsace by the Saint Louis Institute, it was decided to modify the shape of the cube and substitute it with a deeper-shaped piece with an obliquely cut front part.

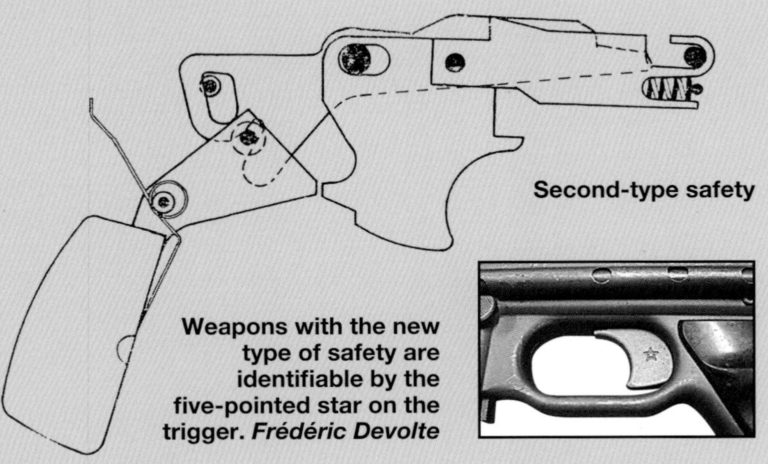

Second-type safety

Weapons with the new type of safety are identifiable by the five-pointed star on the trigger. ***Frédéric Devolte***

Second-type MAT-49 submachine gun. This model has a frame around 5 mm longer compared to the first type, known as "short frame." It is immediately identifiable by the sloping form of the front side of the frame housing, which replaced the "cube" on the previous model.

"Short" frame of the first type (*A*), recognizable by its cube-shaped receiver housing and second-type frame (*B*)

Comparison between the "short" frame of a first-type 49 submachine gun (*C*) and that of a second type (*D*)

In order to compensate for the deeper fixing part of the receiver in the frame, the recoil spring guide was lengthened by 5 mm.

This modification, adopted in May 1950, came into effect during the following months in mass-produced weapons.

This new version of the model 1949 submachine gun, fitted with a frame/receiver part with a sloping side, is generally called "MAT-49 second type" by collectors. It was produced in very large quantities between 1950 and 1962. In addition to the MAT production, an order of 15,000 model 49 submachine guns was made by the Armed Forces Ministry to the weapons manufacturer at Châtellerault on May 11, 1950. It came with the requirement of a total interchangeability of the parts between the weapons produced by both manufacturers. The first submachine guns came out of the MAC in May 1951, and these examples, numbered from A1 to A20, were tested with success by the ETVS. This was already the model 49 second type (with reinforced frame).

The MAC production increased very rapidly: by the end of the year, Châtellerault was producing 200 submachine guns per day. This went up to a rate of 300 per day in 1953 and then stopped at the end of 1954. With the end of the war in Indochina, and taking into account the stocks already built up, the production of the Tulle weapons factory alone was sufficient to satisfy needs at that time. The last batch of twelve weapons, numbered E 019132 to E 019143, was received on November 4, 1954.

In total, out of the more than 500,000 MAT model 1949 second types, only 125,143 were made by the MAC, the remainder being produced by the MAT.

A

B

Stock of the first type (*A*) and second type (*B*), with a reinforcement bracket that also serves as a fixing point for a sling

Position of the magazine folded during maneuvers. With this arrangement, accidental firing is impossible.

Reception of assembled weapons by two military inspectors. *ECPAD*

Later, the MAC continued to supply various spare parts for the model 49 for the MAT up to 1958.

In 1954, a new safety device was tested. It had been observed that since the bolt was hooked on the trigger bar and the safety blocked the trigger, a shot could be fired by pulling the trigger downward. The same incident could be produced by striking the pistol grip.

The trigger and safety mechanisms were therefore modified on all service weapons. The modification was made visible by the marking of a five-pointed star on the trigger.

MAT-49 Third-Type Submachine Guns (Type K)

From 1954 onward, a study of methods of production of the model 1949 was undertaken at the MAT by the lead weapons engineer Bailly and the weapons engineer in chief Deruelle. In parallel with modifications dictated by experience to improve the operational safety of the weapon, the method of production of each part was studied and optimized in terms both of the reduction of costs and the time of production. The replacement of the traditional machining of some parts by precision forging, frittage, and waste-wax casting meant that very precise casts could be produced, which almost always did not require subsequent machining.

View of the underside of the receiver, showing that it is made from a cut, then folded and cast

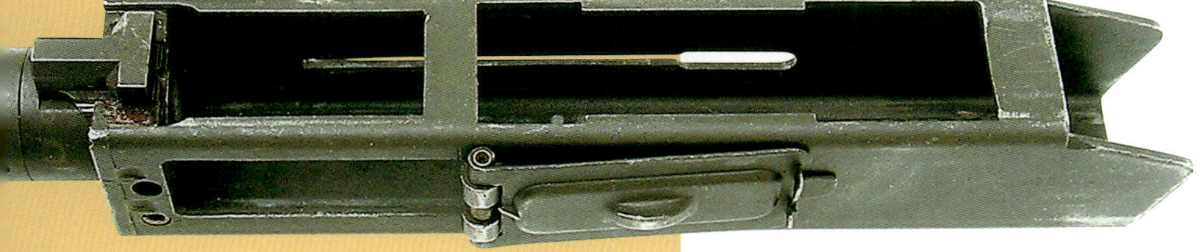

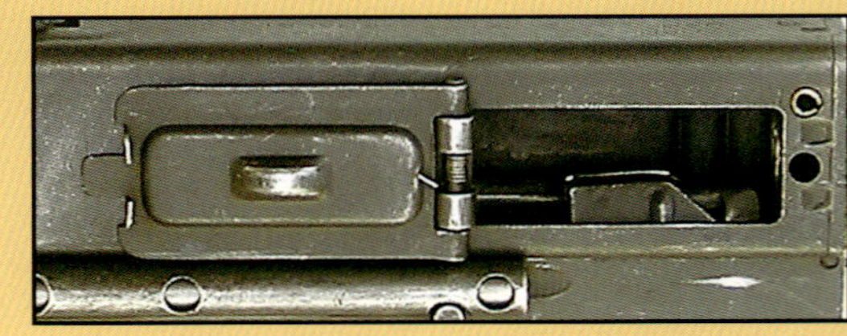

On all versions of the MAT-49 submachine gun, the ejection port is protected by a sealing flap. Under the effect of its spring, the flap opens automatically when the bolt moves rearward when the bolt handle is maneuvered.

Second-type 49 accompanied by a type TAP magazine pouch in canvas (airborne troops). ***Marc de Fromont***

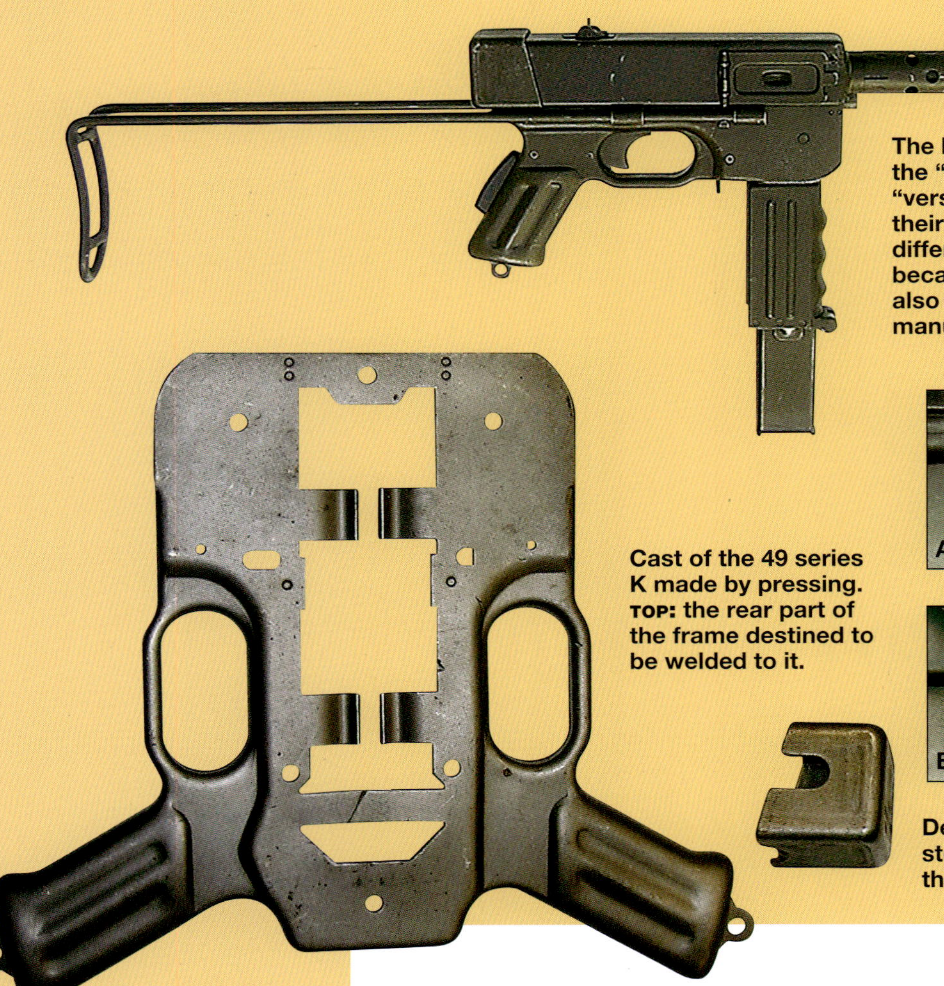

The last variant of the 49, which was christened the "third type" but officially was called "version K" (all the weapons of this series have their serial number preceded by a letter "K"), differs at first glance from the previous ones because its grip is entirely in steel. The weapon also has many simplifications in its manufacture.

Cast of the 49 series K made by pressing. TOP: the rear part of the frame destined to be welded to it.

Development of the guiding tubes on the stock between the second type (*A*) and the third type (*B*)

The results of this study, documented in the memoirs of engineer Bailly ("industrialization of a product—simplification and improvement of mass production, report of an experience"), are very eloquent.

In 1954, it took 582 minutes to make a model 49 submachine gun, and in 1960 just 223 minutes.

In the same year, 450 machine tools to remove shavings were used, as opposed to only eighty-eight in 1960.

Three hundred five minutes of labor were required in 1954 to make a MAT-49, and only seventy hours in 1960, while the cost of production of the weapon was just 64% of the initial price.

In real terms the price of the weapon had been reduced by 45%, with the depreciation of reorganization and the necessary reequipping taken into account in this calculation.

This new method of production transformed, to an extent, the external appearance of the model 49 submachine gun. The most visible manifestation of this development lay in the disappearance of the grip plates in plastic in favor of ones entirely in metal.

This version started to be delivered at the end of the war in Algeria (1962).

At that time, the MAT had made more than 700,000 model 1949 submachine guns (all versions); the MAC had made 125,143. Despite the losses during production and those that were destroyed, at that time around 700,000 MAT-49 submachine guns remained in service.

This was more than was required to equip the French army and the armies of French-speaking Africa, Asia, and the Middle East with whom France had defense and cooperation agreements.

The production of the MAT-49 submachine gun was therefore suspended in Tulle in 1964. Between the first STA and MAS submachine guns of 1924 and the end of the production of the MAT-49, four decades had passed.

Comparison between the receivers of two MAT-49s: *front*, an example of the second type; *back*, an example of the third type (series K)

At the request of the Ministry of the Interior, the MAT developed this long-barrel version of the MAT-49, with a wooden stock and a double trigger permitting single-shot or full-auto fire. Normally the front trigger, controlling full-auto fire, was protected by a shifting flap, and the weapon was used as a semiautomatic carbine. In a crisis situation, the trigger flap was removed and the MAT could then fire on full auto. ***Marc de Fromont***

MAT-49 trigger, protective flap engaged (configuration allowing single-shot firing only) and the trigger flap removed (full-auto fire possible)

After the end of production, the MAT-49 remained in service for another thirty years despite the French army adopting an assault rifle made by FAMAS, the weapons manufacturer at Saint-Etienne, in 1979.

VARIANTS AND MODELS DERIVED FROM THE MAT MODEL 1949 SUBMACHINE GUN

Police Model 49–54

Starting in the 1950s, the national police and the prefecture of Paris police undertook to replace the various submachine guns salvaged at the end of the war (Sten, Thompson, etc.), where their diversity posed problems of maintenance and instruction, with a more homogeneous weapon. The adoption of the MAT model 1949 by the armies rendered an appreciable quantity of MAS model 1938s in 7.65 mm long, which were transferred from the Ministry of Defense to the Ministry of the Interior. At the end of the 1950s, the police also started to replace the MAS-38 with the MAT model 1949 in 9 mm caliber.

In 1951, the technical section of the army carried out tests of a variant of the model 1949 at the request of the Ministry of the Interior; this weapon was developed by the MAT and called "9 mm modified model 49 submachine gun." It was a derivative of the 49 submachine gun, was made by MAT, and had the following particularities:

Marking of the 49-54 police model

- longer barrel (460 mm instead of 230 mm)
- double trigger (single shot and continuous fire, the use of the latter could be made impossible by the use of an eclipsing flap
- wooden stock
- longer sight line (515 mm instead of 367 mm)

The features of this new weapon were as follows: total length, 960 mm instead of 720 mm; total weight, 4.070 kg instead of 3.5 kg.

A MAT-49, or MAT-54, submachine gun was on board every police emergency vehicle. ***Michel Malherbe***

MAT-54 with retractable stock, designed to improve ease of handling of the weapon in police vehicles. ***Michel Malherbe***

MAT-54 SB: This was a version of the model 49 with a modified trigger, allowing only single-shot fire. Only a small quantity of this variant, designed for cash couriers, was made and was never mass-produced.

The objective of the tests led with two modified submachine guns bearing the serial numbers 717 and 720 was the following:

- to compare the ballistic value of the weapon with the 30 M1 carbine
- to compare the wounding power of thirty M1 and 9 mm Parabellum projectiles at 200 meters, this criterion is assessed by shooting at 4 cm thick panels of poplar wood placed 0.5 m one behind the other (at 200 meters, two panels are perforated by 100% of projectiles fired by the M1 carbine, whereas only 80% of projectiles fired by an ordinary 49 submachine gun cross through two panels)
- To examine the repercussions of the modification of the trigger mechanism on operational safety and to verify the resistance during prolonged fire. On the modified submachine guns, after firing 5,216 shots for the first test weapon and 1,152 shots for the second, the single-shot trigger would cease to function due to the deformation of the single-shot trigger spring.
- To study the possible repercussions of any shortening of the barrel on the ballistic value of the weapon. This last point was studied in the possibility of the installation of a fixed folding bayonet on the side of the barrel. It appeared that over the eight lengths of barrel studied (between 460 and 390 mm), the velocity of the projectile at 10 meters and the accuracy at 50 meters in single-shot firing did not vary significantly.

The wounding power of projectiles fired from a modified and a standard model 49 submachine gun with a 230 mm barrel is very similar; however, the accuracy of the modified model was from 20% to 25% higher than that of the standard model.

In conclusion, the STA considered that "the submachine gun derived from the MAT-49 met the demands of its future use very well. Designed to be used as a carbine and offer a generally accurate firing up to 200 meters, because of the presence of a full-auto trigger it can operate full-auto fire in case of emergency, due to having a second trigger. Its dimensions and its wooden stock mean it can be used as a blunt instrument. The presence of a permanently fixed and folding bayonet would make it a complete weapon in terms of the effect on morale that it would confer (nothing would counter the disassembly of the bayonet lug, if trends were to change, the latter being subsequently deemed unnecessary). If the bayonet is not retained, the presence of a 46 cm long barrel on the submachine gun is not justified. Indeed, the weapon conserves all its ballistic qualities with a 38 cm barrel."

Ultimately, the Ministry of the Interior placed an order with MAT for around 800 modified submachine guns with 38 cm barrels, with a wooden stock and two triggers for single shot and full auto. The second of these triggers could be rendered inoperative by the action of a shifting flap. The empty weight of this submachine gun was 4 kg. These police versions, known as the "9 mm Model 1954 submachine gun," were delivered by the MAT during 1955 and 1956.

Third type of MAT-49, with objects evoking the French army in the wake of the war in Algeria. ***Marc de Fromont***

MAT-49 with silencer
TOP: two MAT-49s equipped with ETVS silencer presenting two variants in the surface treatment
BOTTOM: silencer surrounding the barrel formerly developed by an NCO of the 1st Regiment Parachutist Regiment of the Marine Infantry (1st RPIMa).
Marc de Fromont

MAT 49-54 SB
Semiautomatic Carbine

This was a version of a model 1949 submachine gun made in very restricted quantities for the MAT, with the intention of equipping the cash couriers of the bank security services. It had a receiver identical to that of the standard MAT-49 submachine gun, mounted on a frame equipped with a wooden stock and with a safety mechanism allowing only semiautomatic fire. In reality, therefore, it was a semiautomatic carbine rather than a submachine gun.

It would seem that fewer than a hundred examples of this version were produced, the only buyer of which was a single funds transport company, and the Hussard de la Tour du Pin company bought the entire batch around 1990, to be decommissioned and offered for sale to collectors.

MAT Model 1949 with Silencer

Several models of silencer were trialed for the Sten submachine guns, first MAS-38 and then MAT-49. Two organizations took part in the research: the Saint Louis Institute and the ETVS.

Models made within special-forces units, such as the 1st Parachute Regiment of the Marine Infantry, should be added to these official models.

Subsonic 9 mm ammunition, identifiable by the absence of markings on the base, was produced specially for these weapons by French arsenals. Versions of the MAT-49 with silencers were never produced on a large scale, and starting in the 1980s, special forces were equipped with German Hecker & Koch MP 5 submachine guns, some of which were acquired with a silencer (MP 5 SD).

MAT-49 USED FOR INSTRUCTION

Cutaway Weapons

For instructional purposes, some cutaway versions of each of the three versions of the MAT-49 (first, second, and third type) were made, allowing the inner workings of the mechanism to be studied.

Practice Weapons

In parallel, "theoretical" weapons were also made. These weapons were not able to fire and were generally assembled from off-tolerance parts. They were distinguished from standard models by a large letter "X" underlined in white paint, engraved on the

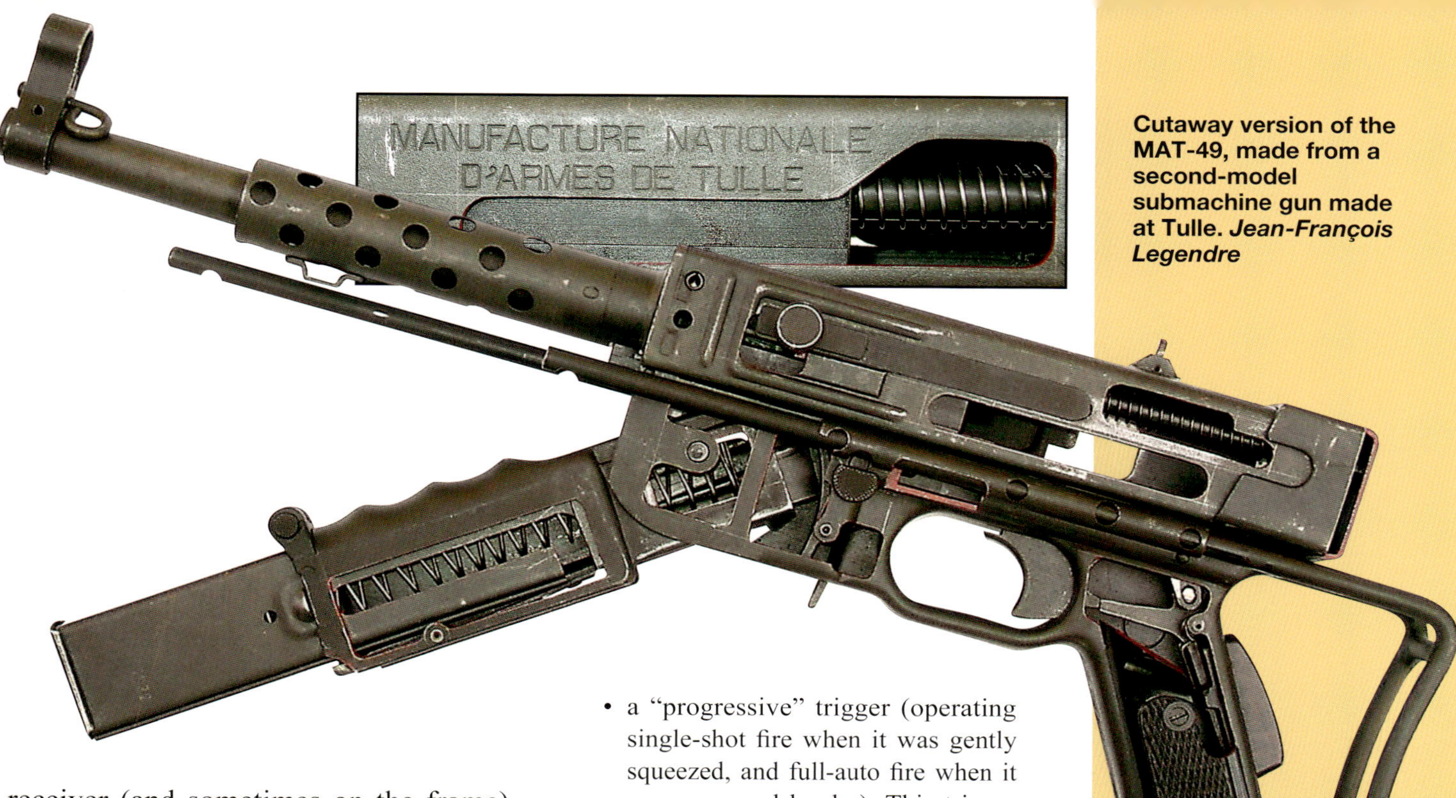

Cutaway version of the MAT-49, made from a second-model submachine gun made at Tulle. ***Jean-François Legendre***

receiver (and sometimes on the frame). Theoretical weapons were placed in training centers to allow recruits to practice assembly and disassembly and handling the weapon without weapons suitable for active service being used.

1.5-Scale Models

As for other light weapons in service in the French army during the 1960s, models of the MAT-49 submachine gun to a scale of 1.5 were also produced in order to present the weapon operation to soldiers during training.

Field-Training Weapons

Weapons that were entirely welded (sometimes called "substitution weapons"), used as a replacement for real ones during obstacle course training, should also be added to these training weapons.

Some MAT-49s were simply neutralized by piercing three holes on the right side of the chamber and the barrel (to avoid any confusion, the zone where these holes were pierced was generally painted in red) and were for a long period used as substitution weapons used for certain types of training (parachuting, nautical exercises). This neutralization removed the necessity of having to apply the same precautions during use and storage as weapons that were able to fire.

Optimized Model

In November 1965, the MAT, which had stopped making the model 49 submachine gun the previous year, made a final attempt to initiate a restart in production by proposing an "optimized" variant of this weapon to the armed forces, with the following features:

- a "progressive" trigger (operating single-shot fire when it was gently squeezed, and full-auto fire when it was squeezed harder). This trigger was also "modular," giving the choice of full-auto or single-shot fire).
- a sight adjustable both in length and direction
- a folding bayonet
- the possibility to mount a grenade launcher

Three cutaway versions of the MAT-49 seen from the left side. TOP TO BOTTOM: first type; second type; third type. ***Marc de Fromon***

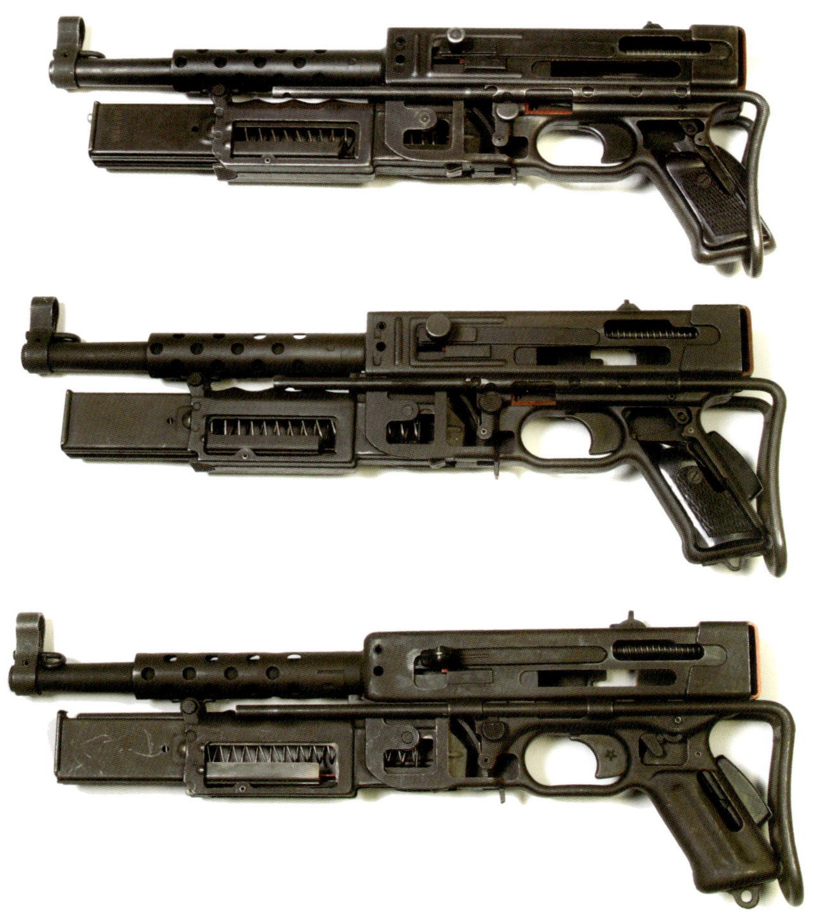

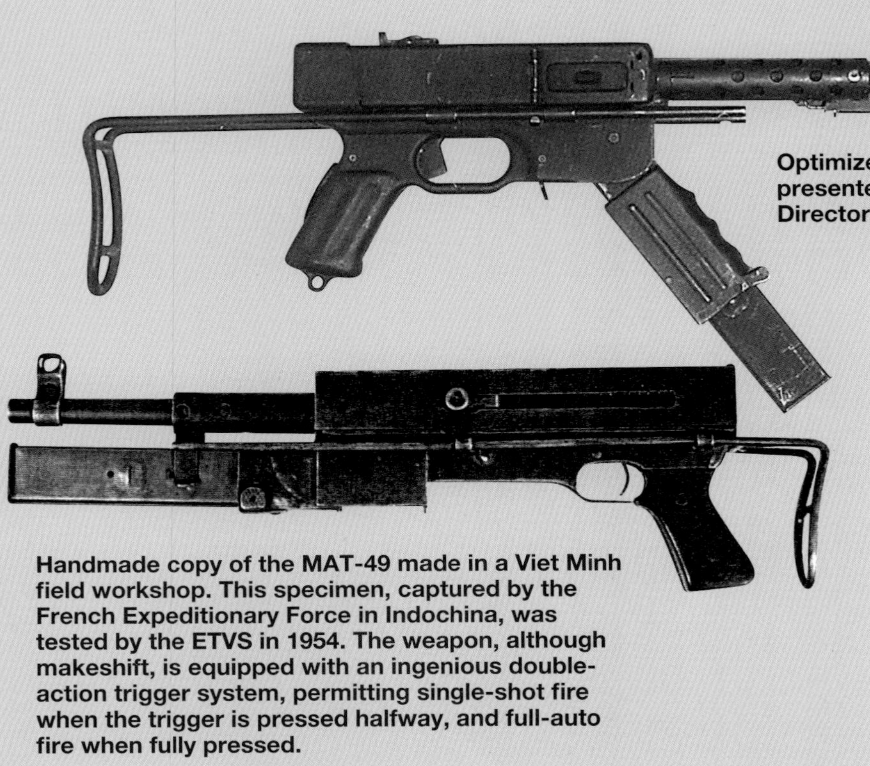

Optimized version of the MAT-49 submachine gun, presented in 1965 by the MAT to the DTAT (Technical Directorate for Land Weapons). ***Jacques Barlerin***

Handmade copy of the MAT-49 made in a Viet Minh field workshop. This specimen, captured by the French Expeditionary Force in Indochina, was tested by the ETVS in 1954. The weapon, although makeshift, is equipped with an ingenious double-action trigger system, permitting single-shot fire when the trigger is pressed halfway, and full-auto fire when fully pressed.

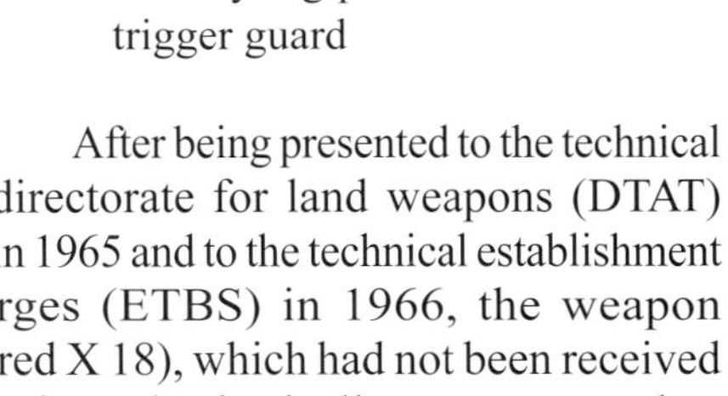

Model 49 submachine gun transformed in 7.62 mm caliber Tokarev. The barrel is significantly longer and thinner than on the original weapon. Note that it is an early-made specimen with a short frame, recognizable by the cube-shaped rear section. ***DR***

- a safety lug positioned in the trigger guard

After being presented to the technical directorate for land weapons (DTAT) in 1965 and to the technical establishment at Bourges (ETBS) in 1966, the weapon (numbered X 18), which had not been received particularly enthusiastically, was returned to the MAT stores before joining those of the Musee de l'Armée several years later.

SUBMACHINE GUNS USED BY THE VIET MINH AND THE VIET CONG

Locally Made Copies

The MAT-49 started its career in Indochina. This new weapon seemed to have been particularly attractive to French adversaries of the time, the Viet Minh fighters, since local copies that had been handcrafted in the field quickly appeared.

A report by the ETVS dated June 15, 1954, relates the assessment of an artisanal copy of the model 49 submachine gun captured from the Viet Minh.

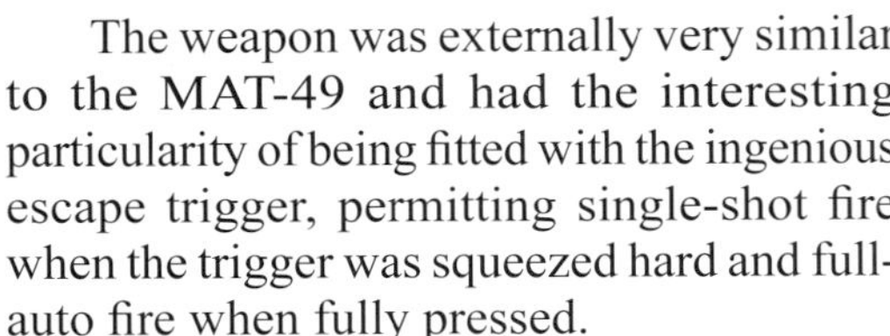

The weapon was externally very similar to the MAT-49 and had the interesting particularity of being fitted with the ingenious escape trigger, permitting single-shot fire when the trigger was squeezed hard and full-auto fire when fully pressed.

The receiver and the grip are made from folded and soldered sheet metal. The handcrafted barrel is, however, rifled; the grip plates are made of horn.

Despite some "firing incidents," with the cartridges stopping the firing of the seventh or eighth cartridge in each burst, five magazines were fired with a mediocre level of accuracy (H + L average: 113 cm) and a high firing speed (880 shots per minute). As the report concludes, "This weapon does not have a great interest but does show, however, that it is possible to make a submachine gun with limited means."

MAT MODEL 49 SUBMACHINE GUN TRANSFORMED INTO 7.62 MM CALIBER TOKAREV

Apart from these handmade productions, a significant number of French 49 submachine guns captured from the French Expeditionary Force or the South Vietnamese army were converted at the end of the 1950s into 7.62 mm caliber Tokarevs, at the time when North Vietnam standardized its armaments alongside Soviet and Chinese norms. Many captured Sten and Thompson submachine guns were also converted into this caliber.

The replacement barrel in 7.62 mm caliber mounted on these weapons is thinner and slightly longer than on the original weapon in 9 mm caliber Parabellum. The magazines appear externally identical to those on the French 49 submachine gun. It is possible that it is a magazine of French origin, with the rib on the rear side removed to allow 7.62 mm Tokarev cartridges to be used, whose total length is greater than that of the 9 mm Parabellum.

Vietcong fighter armed with a captured MAT-49. ***DR***

Police versions: seen here with other objects evoking the events of May 1968: a MAT-49 and a MAT-49-54 submachine gun with their associated equipment in black leather. ***Marc de Fromont***

CHAPTER 11

MAT-49 ACCESSORIES

Twenty-round magazine for desert areas, commonly called a "sand magazine" (*A*), and standard thirty-two-round magazine (*B*)

A B

Using the magazine loader

MAGAZINES AND MAGAZINE LOADERS

Throughout the long career of the MAT-49 submachine gun, its magazines benefited from numerous changes in detail (lengthening of the lips, reinforcement of the recoil spring, improvements in the internal and external surface treatment, etc.).

There were, generally speaking, two main types of magazines: the most standard model containing thirty-two double-stack, double-feed cartridges, and a less common model containing only twenty cartridges on a single stack in order to have a more reliable operation in sandy conditions. For that reason it was known as a "sand magazine."

After various tests on loading tools with spoon lever, a very simple magazine loader based on the version on the American M3 submachine gun was adopted for the MAT-1949 model. This accessory was made from a single piece of thick sheet metal, folded and welded at the rear on the loading hook. There are various types of welding, depending on the period of manufacture.

MAGAZINE POUCHES

During the implementation of the first MAT-49 in the Far East, no specific magazine pouch was provided. The users of these new submachine guns therefore used Sten magazine pouches available on site, or locally made magazine pouches in leather.

The MAT-49 was generally supplied with two magazine pouches, each containing four magazines. The magazine pouches of the initial model had two vertical belt loops at the rear. Later models had a triple fold of leather at the back, allowing the height of the magazine pouch on the belt to be adjusted to suit the shape of the user. A metallic suspension ring allows it to be fixed to the metallic hook on the leather suspenders, supporting the belt.

These magazine pouches came in the standard leather colors of the French army:

- black for the air force, gendarmerie, and law enforcement agencies
- fawn, then subsequently a yellow beige for the other armed forces

View of the most-common models of magazine pouches: fawn leather, beige leather (later model, with three loops to adjust the height), black leather (air force, gendarmerie, and police), TAP-type canvas (airborne troops), 50/53 model (different from the model 50 by the rivets reinforcing the belt suspension loops), and nylon with Velcro fastening (adopted by the air force in the 1990s)

- red for the honor guard of the Spahi units in traditional dress
- a variant in white plastic, also used for the equipment of the honor guard

Magazine pouch chest rig used from the end of the 1950s by marine commandos to carry MAT-49 magazines or antipersonnel rifle grenades

Airborne units were equipped with model 1950 and 1950/53 magazine pouches in canvas, matching the rest of their equipment.

These magazine pouches, worn lower at the belt than leather models, each contained eight magazines. There is a strap under the closing flap, meaning it could be worn over the shoulder.

At the end of the war in Indochina, the marine commandos and some naval detachments operating on land used magazine pouches in canvas and leather with the capacity to carry four or five magazines on the chest. During the war in Algeria, the navy had canvas chest rigs made by its quartermaster, which were reinforced in leather, with five compartments at the rear and front to transport submachine magazines or antipersonnel rifle grenades.

The last type of magazine pouches corresponding to the 1949 model MAT submachine gun appeared at the end of the 1980s, when the air force had olive-green nylon magazine pouches for its air commandos, having a flap closing with a Velcro strip. These magazines were ordered at the same time as the magazine pouches of the same design, but in a smaller size for the two spare magazines accompanying the MAS G1s that the air force had just recently acquired at the time.

In Algiers, during the Algerian war. This parachutist is equipped with a canvas magazine pouch for a submachine gun. *DR*

CLEANING KIT

Apart from the eight magazines, their magazine pouch, a strap, and a user notice, the supply of the MAT model 1949 submachine gun included a leather pouch used to house maintenance accessories. Mention was made of a model 1949 M1A pouch in the list of equipment, the characteristics of which remain unknown.

The most widely used pouch is the 1958 model, which became the model common to all the armed forces from that date onward, after the supplies of the previous model had been used up. This pouch contained the following accessories:

- Model 1949 oil buret in tin-plated metal. This smaller version of the model 1915 oil buret has variations in detail depending on the establishment that made it (MAS, MAT, etc.). Ultimately it went on to be replaced by the 1962 model buret in plastic.
- model 1949 cleaning rod, composed of two sections: a handle and the cleaning part

To take its place in the cleaning pouch, the buret is positioned inside the magazine loader.

Model 1958 pouch for the MAT-49 model 1949 cleaning kit and its contents

Model 1949 oil burets in tin-plated metal and sometimes painted in olive green, marked "Huile" (oil) with black paint, were replaced by the model 1952 buret, in khaki-colored plastic, marked "Armee Française" (French army).

A fawn-colored leather accessory pouch used by the army, and one in black leather used by the gendarmerie

- device for loading the magazines (also known as a magazine loader), positioned around the oil buret

The rags were placed between the different accessories to prevent them from rattling during operations.

This pouch presents the same colors of leather as the magazine pouches. At the end of the war in Indochina, accessory pouches in canvas, able to be used for the transport of MAS-44 or MAS-49 semiautomatic rifle accessories, were sometimes used for submachine gun accessories. It would appear that this use did not last long, and that units using canvas equipment generally went on to use leather cleaning pouches.

SLINGS

A specific sling was created for the MAT-49 submachine gun. This leather sling was in a fawn color for the first makes, whereas later models were in yellow beige. As for magazine pouches and other personal equipment, versions in black leather were ordered by the gendarmerie, police, and air force. Variants in white plastic for honor guards or in red leather for Saharan troops were also made.

Slings in olive-green canvas mounted on 49 submachine guns are occasionally encountered. It seems that these slings were not originally designed to be used for submachine guns but were haversack slings that had been transformed for this use.

BLANK-FIRING ADAPTOR (BFA)

This instrument allows model 1959 plastic blank cartridges to be fired. In 1962, it was decided to reduce the diameter of its vent hole from 3.45 to 3.25 mm to remove the inadvertent bursts of fire linked to an insufficient recoil of the mobile group, to ensure a more regular operation of the weapon and to obtain a firing speed very close to that of real fire.

The same varieties in leather color are seen on the slings as on other equipment: fawn, beige (later), and black, but also white plastic and red leather (not seen in the photo). Canvas slings were used in some units that had canvas equipment; however, this was most likely not a regulation item but rather the adaptation of a haversack sling to the MAT-49 submachine gun in tropical countries, where leather tended to become moldy very quickly. Note that this canvas sling is longer than the leather ones.

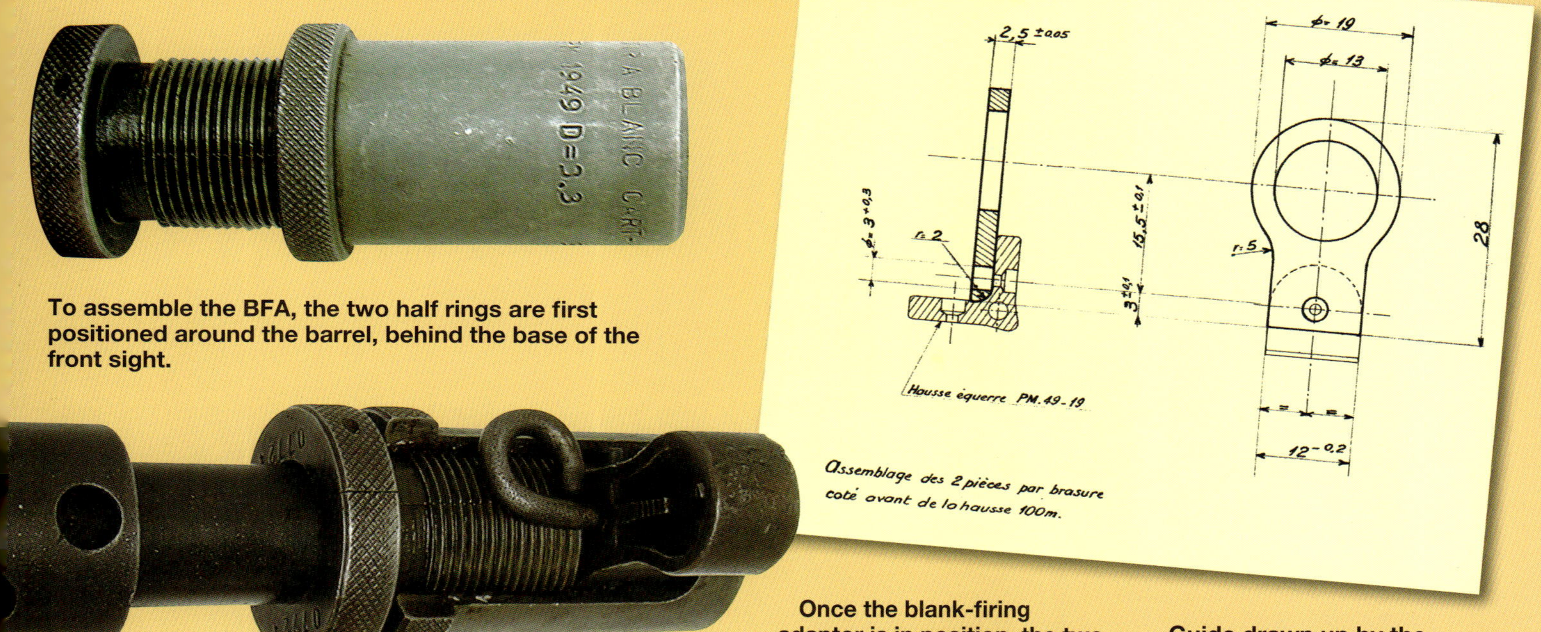

To assemble the BFA, the two half rings are first positioned around the barrel, behind the base of the front sight.

Once the blank-firing adaptor is in position, the two half rings are screwed inside.

Guide drawn up by the test section of the MAS, dated July 26, 1960, for the assembly of tritium pellets on the gunsights on the model 1949 MAT submachine gun. *Jacques Barlerin*

CAMPANA TRAINING DEVICE

This training device, ingeniously designed by Captain Campana, was made for the French army by the Losfeld establishment. It came in a small wooden box and could be adapted to the majority of French light regulation weapons: rifles, semiautomatic rifles, submachine guns, AA52s, and even small-caliber MAS-45 training carbines.

For training with the automatic pistol, the Losfeld establishments also marketed an artificial PA MAC-50 in blue plastic, which allowed for the same type of simulation that the Campana device used with the other types of weapons.

The Campana system was made up of

- a compartment for one battery, which is housed in the magazine well;
- a tube screwed on to the muzzle of the weapon, containing a bulb with a lens to concentrate the beams of light;
- a system of wires and connections to close the electrical circuit and to light the bulb when the user presses the trigger on the weapon; and
- leather slings to secure the wires.

This equipment permits the point of impact to become visible by means of a luminous point on a flat surface, during training in blind-fire or point shooting.

Campana mounted on a MAT-49 model submachine gun

EMPTY-CASE-COLLECTING BAG

This accessory, tested by the infantry group of the Army Technical Section (STA) in 1966, was designed to prevent the task of picking up the cases on the stand after firing. The project came from an Ordnance NCO whose name is unfortunately not mentioned in the test report. The collecting bag was composed of a metal armature with an attached laminated bag that covered the ejection port.

Tests carried out with this bag showed that outside of training on blind-fire or point shooting, the bag did not hinder firing in any way and did not affect the correct operation of the weapon, assuming, of course, that it was regularly emptied.

However, its presence did mean that an extra maneuver, considered dangerous, had to be carried out in order to detach it from the weapon before putting it back in firing mode in the event of a firing incident.

It was for this reason that the empty-case-collecting bag was never issued.

MAS-1967 NIGHT SIGHTS

In order to improve the effectiveness of firing in low-light conditions, tests were carried out to fix luminescent dots on the tritium on MAT-49 sights.

A dot was fixed at the top of a base welded to the rear sight for the distance of 100 meters, and another was welded to the top of the foresight.

This device was never widely used.

CHAPTER 12

AMMUNITION FOR FRENCH POSTWAR SUBMACHINE GUNS

Outline of the 9 mm Parabellum cartridge dated August 28, 1923. Since this cartridge was made only in Germany at that time, its production had been planned for the arsenal at Valence in the event of it being retained as the definitive regulation cartridge of the French army. The abandonment of the 9 mm Parabellum in favor of the 7.65 Long three years later put an end to this project. *Jacques Barlerin*

Three types of ammunition were used for the regulation submachine guns of the French army:

- 7.65 mm Long
- 9 mm Parabellum
- 11.43 mm (.45 ACP)

A detailed study on French-made cartridges for submachine guns alone would be enough for a complete work in itself. In the context of this current study, dedicated to submachine guns of the French army, we will limit ourselves to a basic presentation of the main types of ammunition used by these weapons.

As has already been pointed out in the first chapter, after having established its first submachine gun prototypes in 9 mm caliber Parabellum while waiting for the definitive choice of a cartridge for pistols and submachine guns, the French army opted for the 7.65 mm caliber Long in 1926: ammunition based on the American .30 Pedersen cartridge.

The production of this cartridge was difficult, however. Its use remained anecdotal until the French defeat in 1940, since only a reduced number of pistols and submachine guns that used it had been delivered at the time of the declaration of war.

French-made, postwar 9 mm Parabellum cartridges with steel jacket and ordinary bullet ("O" bullet) with various coverings. *Loïc Helguen*

Under the occupation, the Germans had 7.65 mm Long cartridges with steel casing and a bullet with steel jacket produced by French arsenals.

A decision was made in 1946 by the army general staff to adopt the 9 mm Parabellum cartridge for future pistols and submachine guns of the French army, and as a result, prototypes of submachine guns were tested using German, British, or American cartridges dating from the Second World War. During this time, French cartridge factories began producing 9 mm cartridges, and the arsenals continued making 7.65 mm Long and 11.43 mm (.45 ACP) cartridges while waiting for pistols and submachine guns in 9 mm Parabellum to be in general use.

The first 9 mm Parabellums delivered to the armed forces were cartridges with a phosphate-coated or lacquered cartridge on which projectiles with steel jackets were mounted. These were rapidly abandoned in favor of bullets with brass jackets of various compositions.

Steel cases gave way to brass ones, and after this research phase, production settled on a single model with a brass case and jacket: the model 1950 9 mm cartridge. To ensure a perfectly reliable operation of the submachine gun, this heavily loaded cartridge could cause damage to the model 1950 pistols when they fired a large quantity of bullets (as was the case of supply weapons in training centers, for example).

Only developments in the detail of this cartridge were made until the end of French military production, which came about in 2001 with the closure of the last French cartridge factory working for the army: the GIAT / Matra de Cusset factory.

Apart from ordinary bullet ammunition, the French army also used the following for its 9 mm Parabellum cartridges for its submachine guns:

- tracer bullet (T)
- blank (cartridges with a white plastic body, 59 model)
- dummies
- subsonic ammunition, destined for weapons equipped with silencers

FAMAS F1 and third-type MAT-49 submachine gun. The FAMAS was adopted at the same time as the engagement of French troops in missions of interposition, symbolized here by the uniform of a battalion commander assigned as a United Nations observer. French units of the International Force of the United Nations in Lebanon (FINUL) were for a while equipped with SIG 540 assault rifles while waiting for the supply of the first FAMAS F1s. *Marc de Fromont*

Experimental blank cartridges that resulted in the adoption of the model 1959 blank cartridge. ***Loïc L'Helguen***

Various ammunition in different-colored plastic for blank firing. ***Loïc L'Helguen***

Various French experimental military cartridges. In the middle is a multiple-bullet cartridge. RIGHT: a GIAT cartridge with armor-piercing bullet. ***Loïc L'Helguen***

LEFT: Cartridges with tracer bullets ("T" bullets) with markings in white or red paint or varnish. ***Right*****: a rare example of a tracer bullet in a steel case.** ***Loïc L'Helguen*** **RIGHT: Transport case for 9 mm ammunition boxes, presenting an interesting marking: "reserved for 49 submachine gun—prohibited for use in the Mle 50 and F1 precision model automatic pistols"**

Apart from these standard models, military establishments and civilian cartridge factories researched numerous versions of cartridges of this caliber (with multiple bullets, very high-speed bullets, etc.), which were tested by the armed forces or sometimes even used on a reduced scale, some examples of which are presented here.

TECHNOLOGICAL SURVEILLANCE

Before and after the adoption of the MAT-1949 submachine gun, the technical services of the French army tested the majority of foreign submachine guns in service in Western armies (Madsen, UZI, Vigneron, Sterling, etc.) and various weapons captured by French troops during operations, but also many submachine gun prototypes proposed by French or foreign private companies.

VARIOUS SUBMACHINE GUNS USED BY THE ARMY, THE POLICE, AND GOVERNMENT AGENCIES

Apart from the weapons already presented, other French or foreign submachine guns were (or still are) also used on a smaller scale.

Armed Forces

Rexim Favor: These were weapons captured by the Navy from ships supplying the rebellion during the war in Algeria. They were reused in small quantities by marine commandos. In the 1990s, some of them were still used as substitute weapons during exercises by combat divers from commando Hubert.

Madsen M.50: This machine pistol of Danish origin was used for a certain time by Action Service from the DGSE (Directorate-General for External Security, French secret services) to arm mixed airborne commando groups, operating behind Vietminh enemy lines. The origin of these weapons is not known.

Heckler and Koch MP 5: The diverse variants of which (MP 5SD and MP 5k, etc.) are still used by the special forces.

Heckler and Koch MP 5 A5F and Heckler and Koch UMP: Regulation in the French "Gendarmerie Nationale."

UZI and mini-UZI: Used by some units of engineers (scuba divers and combat engineers from the French 17th Para Engineer Regiment).

Police Forces

UZI machine pistols: Used from the mid-1960s onward by certain police forces (brigades under the direction of the judicial police and regional brigades).

Beretta model 12 S: This currently replaces the MAT model 1949 submachine gun and the UZI submachine gun previously used by police forces.

Heckler and Koch MP 5: Used by the Research and Intervention Brigade of the French Interior Ministry and the Paris criminal police.

The Ministry of the Interior also tested Walther MPL and MPK submachine guns in the 1970s. The specimens presented for these tests bear the marking of the Manurhin company.

Beretta 12 S model: Currently used by the French national police.

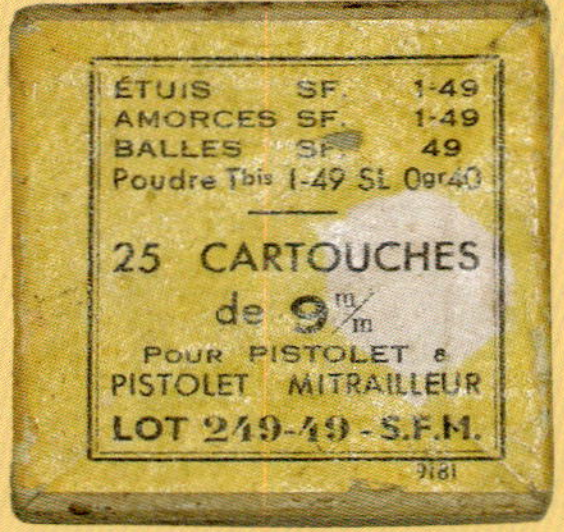

Box of twenty-five cartridges of 9 mm Parabellum for machine pistols made by the Saint Louis Institute in 1949

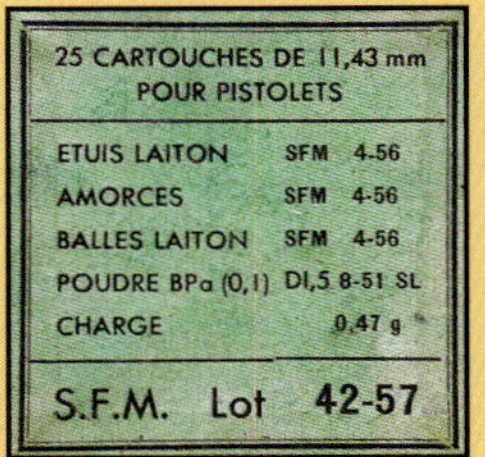

Box of French military–made 11.43 mm (.45 ACP) cartridges. ***Loïc L'Helguen***

Favor version with a bayonet and a protective barrel jacket

The UZI submachine gun: this weapon, of Israeli origin, was also made under license in Belgium by the FN Herstal and was used by many sections of the French police.

CLOSE-DEFENSE WEAPON

The close-defense weapon takes up the idea of the small-size submachine gun, designed to arm soldiers whose main task does not involve using an individual weapon (NCOs, drivers, signalmen, et al.), thereby giving them a weapon that was compact but with great firepower, allowing them to defend themselves in crisis situations.

A similar project had already been carried out in the 1950s by the MAS with the model 55 MAS submachine gun: a small weapon, in 9 mm caliber, equipped with a stock holster. This project was not completed.

Starting in 1987, a new type of close-defense weapon was researched by the equipment-manufacturing facility at Bourges (EFAB/GIAT Bourges). Initially christened PAD (self-defense pistol) and then AAD (self-defense weapon), this initial version fired a bottleneck cartridge of 5.56 × 25 caliber.

This version, which was presented to the specialized press in 1988, did not give total satisfaction. This project was taken up in 1992 by the weapons and ammunitions research department of the national weapons factory at Saint-Etienne, where it was the last research carried out.

The development of a new ammunition designed for this weapon was entrusted to the cartridge factory at Le Mans. This new ammunition kept the bottleneck shape of the 5.56 × 25 mm, but its cartridge was 3 mm shorter. This relatively short (5.56 × 22 mm) cartridge had the advantage of being able to be used to transform, without great difficulty in this caliber, weapons initially chambered in 9 mm Parabellum (such as the MAS G1 pistol, for example).

The 5.56 × 22 mm cartridge, whose development was finalized in 1990, proved to be capable to penetrate all bulletproof vests of the period at a short distance.

The takeover of the Belgian factory FN Herstal by GIAT forced the close-defense-weapon project to be abandoned in 1994 in favor of the P.90, developed by the FN Herstal.

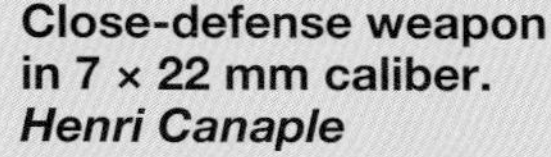

9 mm caliber Parabellum submachine gun presented by the Manurhin company in the 1980s to the Minister of the Interior. *Jean François Legendre*

Close-defense weapon in 7 × 22 mm caliber. *Henri Canaple*

Alain Delon in the film *Les Centurions*

The 8th RCP in Algeria

CONCLUSION

In 1979, the assault rifle of the weapons factory of Saint Etienne, the FAMAS, was adopted by the French army to replace the rifles and submachine guns in service. The replacement of the MAT-49 submachine gun by the FAMAS took place gradually, first and foremost in the land army and naval fusilier units, and then the air force.

The MAT-49 weapons, gradually withdrawn from active service, were initially placed in reserve in the event of mobilization. They appeared on occasion to arm reserve units until recently.

Several of these weapons were neutralized for the benefit of collectors, others were transformed into field-training weapons, and the majority were gradually destroyed.

The MAT-49 was both well designed and well made and left all those who used it with the memory of a reliable, solid, and easily transportable weapon that was easy to operate and maintain.

ACKNOWLEDGMENTS

Thanks to the personnel of the Gendarmerie Nationale, the Ministry of the Interior, and the archives center of the Arms Manufacturer at Châtellerault. Thanks to Marc de Fromont for taking the full-page photos, as well as Jacques Barlerin, Michel Moreau, Eric Klamerek, Patrick Walter, Loïc L'Helguen, Wolf Reiss, and Michel Malherbe.

Jacques Barlerin, who brought his invaluable help to the creation of this book, died suddenly on February 12, 2007, at the age of sixty-four. It was with great sadness that we learned of the death of our kind and humorous friend. His exceptional knowledge of French weapons and ammunition manufacture had its source in his vast network of personal contacts within the French armaments industry, and in the long years of research undertaken in the archives of arms manufacturers and the archive center at Châtellerault. His death represents an irreparable loss for all those interested in the history of French armaments.

BIBLIOGRAPHY

Articles

Barlerin, Jacques. "Un accessoire hors du commun: La baïonnette du PM MAT 49." *La Gazette des Armes* 338.

Barlerin, Jacques. "Les cartouches de 9 mm Parabellum en France." *Calibres*, October 2006.

Druard, Michel, and René Smeets. "Sola Super, Sola Légère et mini-Sola." AMI.

Duquesne, Michel J. H. "La mitraillette RAN, concurrente malheureuse de la Vigneron." AMI.

Ferrard, Stéphane. "Le premier PM français réglementaire: Le STA calibre 9 mm." *La Gazette des Armes* 65.

Fichant, Jean Michel. "Les pistolets-mitrailleurs Hotchkiss de l'après-guerre." AMI.

Huon, Jean. "Les armes automatiques MGD." *La Gazette des Armes*.

Huon, Jean. "Les pistolets-mitrailleurs Hotchkiss." *Cibles* 118.

Malherbe, Michel. "La carabine mitrailleuse Hotchkiss (CHM2) type universel." *La Gazette des Armes* 321.

Mastier, Dominique, and Stéphane Ferrard. "Le chemin de croix du PM MAS 38." *La Gazette des Armes* 66.

Regenstreig, Philippe. "Le pistolet mitrailleur Delacre." *La Gazette des Armes* 314.

Vauvillier, François. "Mitraillettes et Bourguignottes." *La Gazette des Armes* 31 and 32 (October and November 1975).

Books

Malherbe, Michel. *Les armes de la police nationale*. Jacques Grancher, 1983.

Malherbe, Michel. *Les pistolets-mitrailleurs européens*. Editions ELT, 1985.

Merglen, Albert. *Groupe Franc*. Arthaud, 1943.

Moreau, Michel. *La mitraillette STEN en France*. Unpublished case study.

Saint-Roc. *Sacrée drôle de guerre*. Paris: Edition Nouveau Parthenon, 1948.

CLASSIC GUNS OF THE WORLD SERIES